ENERGY POLICIES, POLITICS AND PRICES

DEEPWATER HORIZON

FEDERAL FINANCIAL RISKS FROM OIL SPILLS

ENERGY POLICIES, POLITICS AND PRICES

Additional books in this series can be found on Nova's website under the Series tab.

Additional E-books in this series can be found on Nova's website under the E-book tab.

ECONOMIC ISSUES, PROBLEMS AND PERSPECTIVES

Additional books in this series can be found on Nova's website under the Series tab.

Additional E-books in this series can be found on Nova's website under the E-book tab.

ENERGY POLICIES, POLITICS AND PRICES

DEEPWATER HORIZON

FEDERAL FINANCIAL RISKS FROM OIL SPILLS

Denis A. Michaels
and
Gregorio S. Aston
EDITORS

Nova Science Publishers, Inc.
New York

For permission to use material from this book please contact us:
Telephone 631-231-7269; Fax 631-231-8175
Web Site: http://www.novapublishers.com

Library of Congress Cataloging-in-Publication Data

ISBN 978-1-61942-499-9

Published by Nova Science Publishers, Inc. † New York

CONTENTS

PREFACE

Both the individual circumstances of the Deepwater Horizon incident, as well as the overall framework for how the federal government responds to oil spills, present a mix of evolving, but as yet uncertain, financial risks to the federal government and its Oil Spill Liability Trust Fund. The extent of financial risks to the federal government from the Deepwater Horizon is closely tied to BP and the other responsible parties. This new book identifies the financial risks to the federal government resulting from the Deepwater Horizon oil spill; and assesses the Coast Guard's internal controls for ensuring that processes and payments for spill-related cost reimbursements and claims related to the spill are appropriate. Also discussed is the extent to which the federal government oversees the BP and Gulf Coast Claims Facility cost reimbursement and claims processes

Chapter 1- On April 20, 2010, an explosion of the *Deepwater Horizon* oil rig leased by BP America Production Company (BP) resulted in a significant oil spill. GAO was requested to (1) identify the financial risks to the federal government resulting from oil spills, particularly *Deepwater Horizon, (2)* assess the Coast Guard's internal controls for ensuring that processes and payments for spill-related cost reimbursements and claims related to the spill are appropriate, and (3) describe the extent to which the federal government oversees the BP and Gulf Coast Claims Facility cost reimbursement and claims processes. We issued status reports in November 2010 and April 2011. This is the third and final report related to these objectives. We obtained and analyzed data on costs incurred from April 2010 through May 2011 and claims submitted and processed from September 2010 through May 2011. We reviewed relevant policies and procedures, interviewed officials and staff at key federal departments and agencies, and tested a sample of claims processed and cost reimbursements paid for compliance with internal controls.

Chapter 2- On April 20, 2010, an explosion occurred on BP America Production Company's (BP) leased mobile offshore drilling unit *Deepwater Horizon*. The total cost to clean up the massive and unprecedented oil spill in the Gulf of Mexico following the *Deepwater Horizon* explosion (including costs to help pay for the spill's adverse impact on businesses and individuals in the region) are yet unknown, but have been estimated in the tens of billions of dollars. The extent to which the federal government will ultimately be required to pay costs associated with the *Deepwater Horizon* oil spill remains unclear.

The complex legal framework in place for oil spill liability and response funding will play an integral role in determining who is responsible and will ultimately pay the costs associated with the *Deepwater Horizon* oil spill. In this regard, the Oil Pollution Act of 1990,[1] as amended (OPA), which Congress enacted after the *Exxon Valdez* spill in 1989, authorized use of the Oil Spill Liability Trust Fund (Fund) to pay for certain oil spill cleanup costs and damages using federal tax revenues for immediate response costs and when the responsible parties cannot be identified or do not pay. OPA also provided that the federal government may subsequently seek reimbursement for these costs from responsible parties.[2] The Fund, which is administered by Coast Guard's National Pollution Funds Center (NPFC), is subject to a $1-billion cap on the total amount of expenditures per incident.

Chapter 3- On April 20, 2010, an oil spill of national significance in the Gulf of Mexico followed an explosion on the mobile offshore drilling unit *Deepwater Horizon* (the *Deepwater Horizon* oil spill). The *Deepwater Horizon* was leased by BP America Production Company (BP) as part of the Macondo project.[1] 152 days later, on September 19, 2010, BP confirmed the completion of cementing operations to prevent further oil from spilling from the Macondo Prospect well to which the *Deepwater Horizon* was attached when it exploded. In order to coordinate the federal response to the *Deepwater Horizon* oil spill, the National Incident Commander established the Deepwater Integrated Services Team (IST) consisting of 18 federal agencies, including the U.S. Coast Guard and the Department of Justice (DOJ).

In Deepwater Horizon: ISBN: 978-1- 61942-499-9
Editors: D. A. Michaels and G. S. Aston © 2012 Nova Science Publishers, Inc

Chapter 1

DEEPWATER HORIZON OIL SPILL: ACTIONS NEEDED TO REDUCE EVOLVING BUT UNCERTAIN FEDERAL FINANCIAL RISKS*

United States Government Accountability Office

WHY GAO DID THIS STUDY

On April 20, 2010, an explosion of the *Deepwater Horizon* oil rig leased by BP America Production Company (BP) resulted in a significant oil spill. GAO was requested to (1) identify the financial risks to the federal government resulting from oil spills, particularly *Deepwater Horizon, (2)* assess the Coast Guard's internal controls for ensuring that processes and payments for spill-related cost reimbursements and claims related to the spill are appropriate, and (3) describe the extent to which the federal government oversees the BP and Gulf Coast Claims Facility cost reimbursement and claims processes. We issued status reports in November 2010 and April 2011. This is the third and final report related to these objectives. We obtained and analyzed data on costs incurred from April 2010 through May 2011 and claims

* This is an edited, reformatted and augmented version of the United States Government Accountability Office publication, Report to Congressional Requesters, GAO-12-86, dated October 2011.

submitted and processed from September 2010 through May 2011. We reviewed relevant policies and procedures, interviewed officials and staff at key federal departments and agencies, and tested a sample of claims processed and cost reimbursements paid for compliance with internal controls.

What GAO Recommends

GAO is (1) reiterating that Congress may want to consider setting a Fund cap per incident based upon net expenditures, (2) presenting a new matter concerning extending the barrel tax used to finance federal oil spill responses to sustain program funding, and (3) making a recommendation to improve procedures for future significant spills. In responding, the Department of Homeland Security concurred with the recommendation.

What GAO Found

Both the individual circumstances of the *Deepwater Horizon* incident, as well as the overall framework for how the federal government responds to oil spills, present a mix of evolving, but as yet uncertain, financial risks to the federal government and its Oil Spill Liability Trust Fund (Fund). The extent of financial risks to the federal government from the *Deepwater Horizon* is closely tied to BP and the other responsible parties. BP established a $20 billion Trust to pay for individual and business claims and other expenses. As of May 31, 2011, BP has paid over $700 million of federal and state government costs for oil spill cleanup. Federal agency cleanup and restoration activities are under way and agencies continue to incur costs and submit them for reimbursement. However, the full extent of these costs, particularly those related to environmental cleanup, may not be fully realized for some time. As cleanup costs continue to mount, it is possible that expenditures from the Fund will reach the $1 billion total expenditure per incident cap. Expenditures were over $626 million on May 31, 2011. If these amounts reach the total expenditure cap of $1 billion, the Fund can no longer be used to make payments to reimburse agencies' costs (or to pay valid individual or business claims if not paid by the responsible parties). At that point, government agencies would no longer be able to obtain reimbursement for their costs. In November 2010, GAO suggested that Congress may want to consider setting a

Fund per incident cap based on net expenditures (expenditures less reimbursement), rather than total expenditures. Finally, GAO found the federal government's longer-term ability to provide financial support in response to future oil spills is also at risk because the Fund's primary source of revenue, a tax on petroleum products, is scheduled to expire in 2017.

GAO's testing of the Coast Guard's internal controls over *Deepwater Horizon* claims processed and cost reimbursements processed and paid showed that adjudicated claims processed and costs reimbursed were appropriate and properly documented. In November 2010, GAO made four recommendations regarding establishing and maintaining effective cost reimbursement policies and procedures for the Fund. The Coast Guard changed its operating practices to reflect lessons learned from the initial response to the *Deepwater Horizon* incident, and it has updated its cost reimbursement procedures accordingly. However, the Coast Guard has not yet updated its procedures for processing significant claims, so lessons learned from its experiences processing *Deepwater Horizon* claims could be lost. Capturing lessons learned about processing such claims will be essential should a significant spill occur in the future.

The federal government has used a variety of approaches to oversee BP's and GCCF's cost reimbursement and claims processing. Soon after the *Deepwater Horizon* oil spill, the federal government established a Deepwater Integrated Services Team (IST), which was initially responsible for monitoring BP's claims process, among other things. Subsequently, the oversight of cost reimbursement and claims activities transitioned to the Department of Justice, which continues to lead this and other efforts. In addition, the Department of the Interior and the National Oceanic and Atmospheric Administration are serving as the federal government's representatives for the natural resource trustees in evaluating the environmental impact of the *Deepwater Horizon* spill and selecting and implementing restoration projects to be funded by BP.

ABBREVIATIONS

BP	BP American Production Company
COFR	Certificate of Financial Responsibility
DHS	Department of Homeland Security
DOD	Department of Defense
DOI	Department of the Interior
EPA	Environmental Protection Agency

FEMA	Federal Emergency Management Agency
FINCEN	Coast Guard's Finance Center
FOSC	Federal On-Scene Coordinator
GCCF	Gulf Coast Claims Facility
IG	inspector general
IST	Integrated Services Team
MIPR	Military Interdepartmental Purchase Request
NOAA	National Oceanic and Atmospheric Administration
NPFC	National Pollution Funds Center
OMB	Office of Management and Budget
OPA	Oil Pollution Act of 1990, as amended
PRFA	Pollution Removal Funding Authorization

October 24, 2011
Congressional Requesters

The explosion on the BP America Production Company's (BP) leased *Deepwater Horizon* oil rig in the Gulf of Mexico on April 20, 2010 resulted in one of the largest environmental disasters in U.S. history.[1] Along with the devastating environmental impact, the *Deepwater Horizon* oil spill affected the livelihoods of thousands of Gulf Coast citizens and businesses. The total costs to clean up this unprecedented spill, ease the economic suffering of affected parties in the region, and assess and mitigate its eventual environmental impact remain unknown but have been estimated in the tens of billions of dollars.

The Oil Pollution Act of 1990, as amended (OPA), which was enacted after the *Exxon Valdez* oil spill in 1989, established a "polluter pays" system that places the primary burden of liability for costs of spills up to a statutory maximum, on the responsible parties—BP and several other companies in this case.[2] However, responsible parties are liable without limit if the oil discharge is the result of gross negligence or willful misconduct, or a violation of federal operation, safety, and construction regulations. OPA provides the Oil Spill Liability Trust Fund (Fund) to pay for oil spill costs when the responsible party cannot or does not pay. The Fund is administered by the Coast Guard through its National Pollution Funds Center (NPFC), is primarily financed through a tax on petroleum products, and is subject to a $1 billion cap on the amount of expenditures from the Fund per incident.[3] NPFC administers the Fund by disbursing funds to government agencies to reimburse them for their oil spill cleanup costs (cost reimbursements), monitoring the sources and uses of funds, adjudicating claims submitted by individuals and businesses to the Fund for payment (claims), and pursuing reimbursement from the responsible party for costs and damages paid from the Fund (billing the responsible party).

In the case of the *Deepwater Horizon* oil spill, BP, as well as others, have been identified as responsible parties. In this capacity, BP established multiple claims centers along the Gulf Coast to receive and process individuals' and businesses' damage claims; on May 3, 2010, BP began paying emergency compensation to them. In June 2010, as part of an oral agreement between the Obama Administration and BP, BP established a new claims processing facility, the Gulf Coast Claims Facility (GCCF). GCCF, administered by Kenneth Feinberg, began operations on August 23, 2010, and is responsible for handling claims from individuals and businesses for damages resulting from the spill. BP also established an irrevocable trust in August 2010—primarily to pay claims approved by GCCF among other purposes—and pledged to incrementally provide a total of $20 billion to the trust by 2014.[4]

Shortly after the April 20, 2010 explosion, members of Congress requested that we (1) identify the financial risks to the federal government resulting from oil spills, particularly *Deepwater Horizon*, (2) assess NPFC's internal controls for ensuring that processes and payments for cost reimbursements and processes for claims related to the *Deepwater Horizon* spill were appropriate, and (3) describe the extent to which the federal government oversees the BP and GCCF *Deepwater Horizon* oil spill-related claims processes.

This report is the third and final in a series of reports related to these objectives. In November 2010, we provided our preliminary assessment of the financial risks and the cost reimbursement and notification policies and procedures associated with the *Deepwater Horizon* oil spill.[5] We expressed the view that Congress should consider changing the calculation of expenditures made against the Fund's $1 billion per incident expenditure cap to take into account reimbursements from responsible parties. We also made four recommendations to NPFC directed at helping NPFC establish and maintain effective cost reimbursement policies and procedures for the Fund and update NPFC's current policies to reflect current organization, structure, and management directives. (See appendix I for the specific recommendations and their status.) In April 2011, we provided updated information on the financial risks to the federal government associated with the Fund's cap on total, rather than net, expenditures, as well as claims submitted to and reviewed by NPFC and GCCF.[6] This final report provides an update on the issues and risks to the federal government, the results of our testing of NPFC's internal controls over the processes and payments for cost reimbursement and the claims process, and an overview of the federal government's oversight of BP and GCCF's claims processes.

As part of our analysis of financial risks to the federal government and the Fund, we identified and analyzed applicable laws and regulations to determine statutory and regulatory limitations on the liability of responsible parties that may pose financial risks to the Fund and federal government. As one of OPA's goals is to make the environment and public whole for injuries to natural resources resulting from an oil spill, we reviewed applicable guidance, regulations, and NPFC annual reports to gain an understanding of the Natural Resource Damage Assessment[7] process. These processes involve determining the type and amount of restoration needed to compensate the public for harm to natural resources as a result of an oil spill, and the length of time these assessments may take to complete. In addition, we reviewed publicly available quarterly financial information of responsible parties through June 2011 to gain an understanding of the extent to which these companies reported contingent liabilities.[8] To determine the amounts obligated and actual costs incurred in relation to the Fund's $1 billion per incident cap, we obtained and analyzed daily financial summary data NPFC uses to track costs for the *Deepwater Horizon* oil spill. We obtained invoices NPFC sent to the responsible parties to reimburse the Fund, analyzed the requests for reimbursements submitted by federal and state agencies, and compared the invoiced amounts to the amounts federal and state agencies had submitted for payment from the Fund.

To assess the extent to which NPFC's internal controls ensured that cost reimbursements were appropriate, we tested a statistical sample of payments made to federal and state agencies between April 2010 and April 2011 for *Deepwater Horizon* removal and response activities for compliance with NPFC's policies and procedures. In addition, while NPFC had not made any payments in response to claims from individuals and businesses submitted as of April 30, 2011, we tested a statistical sample of *Deepwater Horizon* final claim determinations—all denials or withdrawn by the claimant—that had been made by NPFC officials for compliance with NPFC's policies and procedures to implement OPA requirements. We also obtained information on NPFC's claims contingency planning for handling a potentially large number of claims related to the *Deepwater Horizon* oil spill. In order to update information about claims submitted and reviewed by NPFC and GCCF, we used available NPFC and GCCF claims data through May 31, 2011.

To provide an overview of the extent to which the federal government oversees the BP and the GCCF claims processes, we interviewed agency officials about oversight of BP's claims process. Specifically, we interviewed Department of Justice (Justice) officials who worked with BP to establish the GCCF and its claims processes, among other things. We also interviewed NPFC officials who monitor GCCF's actions to approve and deny claims from individuals and businesses. We interviewed Justice and Office of Management and Budget (OMB) officials about any plans to pursue payment from the responsible parties for federal government costs for the *Deepwater Horizon* oil spill that are not reimbursed through an intragovernmental agency agreement. Appendix II provides additional details on our scope and methodology.

We conducted this performance audit from July 2010 to October 2011, in accordance with generally accepted government auditing standards. Those standards require that we plan and perform the audit to obtain sufficient, appropriate evidence to provide a reasonable basis for our findings and conclusions based on our audit objectives. We believe that the evidence obtained provides a reasonable basis for our findings and conclusions based on our audit objectives.

BACKGROUND

Legal Framework Establishing Responsibilities to Pay Oil Spill Costs and Claims

The legal framework for addressing and paying for maritime oil spills is identified in OPA, which was enacted after the 1989 *Exxon Valdez* spill. OPA places the primary burden of liability and the costs of oil spills on the owner and operator of the vessel or on shore facility and the lessee or permittee of the area in which an offshore facility is located. This "polluter pays" framework requires that the responsible party or parties assume the burden of spill response, natural resource restoration, and compensation to those damaged by the spill, up to a specified limit of liability. In general, the level of potential exposure under OPA depends on the kind of vessel or facility from which a spill originates and is limited in amount unless the oil discharge is the result of gross negligence or willful misconduct, or a violation of federal operation, safety, and construction regulations, in which case liability under OPA is unlimited.[9] For oil spills from an offshore facility, such as the *Deepwater*

Horizon, liability is limited to all removal—or cleanup—costs plus $75 million.[10]

Under OPA, before any vessel larger than 300 gross tons can operate in U.S. waters, the owner/operator must obtain a Certificate of Financial Responsibility (COFR) from NPFC. This COFR demonstrates that the owner/operator has provided evidence of financial responsibility to pay for removal costs and damages up to the liability limits required by OPA. These OPA requirements for demonstrating financial responsibility apply only to the statutory maximum amount of potential liability under OPA, although states may impose additional liabilities and requirements related to oil spills in state waters.[11]

OPA requires that, subject to certain exceptions, such as removal cost claims by states, all nonfederal claims for OPA-compensable removal or damages be submitted first to the responsible party or the responsible party's guarantor. If the responsible party denies a claim or does not settle it within 90 days, a claimant may present the claim to the federal government to be considered for payment.[12] To pay specified claims above a responsible party's liability limit, as well as to pay claims when a responsible party does not pay or cannot be identified, OPA authorizes use of the Fund subject to limitations on the amount and types of costs. For example, under OPA, the authorized limit on Fund expenditures for a single spill is currently set at $1 billion (without consideration of whether the Fund was reimbursed for any expenditures). In addition to paying claims, the Fund is used to reimburse government agencies for certain eligible costs they incur. Further, within the $1 billion cap, the costs for conducting a natural resource damage assessment and claims paid in connection with any single incident shall not exceed $500 million. OPA provides that the President designate the federal officials and that the governors designate the state and local officials who act on behalf of the public as trustees for natural resources.[13] OPA regulations provide that the trustees may recover costs for natural resource damage assessment and restoration.[14] The Fund may not be used for certain types of personal injuries or damages that may arise related to an oil spill incident, such as financial losses associated with oil company investments by members of the public. Recovery for such damages and injuries may be governed by other federal statutes, common law, or various state laws.

Federal agencies are authorized to use the Fund to cover their oil removal costs from the affected areas to the extent the Fund has funds available within the $1 billion cap.[15] The federal government is entitled to reimbursement from responsible parties for such costs.

The Coast Guard's NPFC administers uses of the Fund to reimburse government agencies for their removal and cleanup costs;[16] adjudicating individual and business claims submitted to the Fund for payment; and pursuing reimbursement from the responsible party for costs and claims paid by the Fund. NPFC bills the responsible parties directly, including BP in this case, for costs government agencies have incurred, and all payments received from responsible parties are deposited into the Fund.

OPA defines the costs for which responsible parties are liable and for which the Fund is made available for compensation in the event that the responsible party does not pay, cannot pay, or is not identified. As described in greater detail in appendix V, "OPA compensable" costs include two main types:

- Removal Costs: Removal costs are incurred by the federal government or any other entity taking approved action to respond to, contain, and clean up the spill. For example, removal costs include cleaning up adjoining shoreline affected by the oil spill and the equipment used in the response—skimmers to pull oil from the water, booms to contain the oil, planes for aerial observation—as well as salaries, travel, and lodging costs for responders.
- Damages: OPA-compensable damages cover a wide range of both actual and potential adverse impacts from an oil spill. For example, damages from an oil spill include the loss of profits to the owner of a commercial charter boat if the boat was trapped in port because the Coast Guard closed the waterway in order to remove the oil, or personal property damage to the owner of a recreational boat or waterfront property that was oiled by the spill, for which a claim may be made first to the responsible party, if possible, or to the Fund.

In addition to OPA-compensable costs, the federal government can also incur other non OPA-compensable costs associated with oil spills. For example, the federal government had various non-OPA-compensable costs for the *Deepwater Horizon* oil spill, such as Department of Homeland Security (DHS) costs associated with providing additional staff to NPFC for receiving and processing claims.[17]

Four Operational Response Phases for Oil Removal

The *National Oil and Hazardous Substances Pollution Contingency Plan*, more commonly called the National Contingency Plan is the federal government's blueprint for responding to oil spill and hazardous substance releases. The National Contingency Plan provides the organizational structure and procedures for preparing for and responding to discharges of oil and releases of hazardous substances, pollutants, and contaminants.[18] The plan outlines approved procedures and removal activities when responding to an oil spill and identifies the following four phases of response operations for oil discharges:

1) Discovery and Notification include activities conducted to discover oil spills or to notify appropriate authorities of oil spills.
2) Preliminary Assessment and Initiation of Action include activities conducted to assess the magnitude and severity of the spill and to assess the feasibility of removal and plan appropriate actions. These activities are necessary whether or not the responsible party is taking action.
3) Containment, Countermeasures, Cleanup, and Disposal include oil spill cleanup activities such as hiring contractors and transporting and staging required supplies and needed equipment.
4) Documentation and Cost Recovery include the activities necessary to support cost recovery and record uses of the Fund.

Three of the four phases for oil removal remain under way for the *Deepwater Horizon* incident, and the operational response is likely to continue for years. The first phase, discovery and notification, is substantially complete. Subject to certain thresholds, the costs incurred in phases two, three, and four are eligible to be paid from the Fund.

The Fund's Financial Resources to Pay Oil Spill Costs and Claims

The Fund's primary revenue source is an 8 cent per barrel tax on petroleum products either produced in the United States or imported from other countries. Other revenue sources include recoveries from responsible parties for costs of removal and damages, fines and penalties paid pursuant to various statutes, and interest earned on the Fund's U.S. Treasury investments. In fiscal year 2009, the barrel tax was 92 percent of the Fund's revenue. As shown in figure 1, the Fund's balance has varied over the years. The barrel tax expired in December 1994 and was reinstituted at 5 cents per barrel in April 2006 as mandated by the Energy Policy Act of 2005. The Energy Improvement and Extension Act of 2008 increased the tax to 8 cents per barrel and provides that the Fund's barrel tax shall expire after December 31, 2017.[19]

In fiscal year 2011, the increase to the Fund is primarily attributable to reimbursements received from responsible parties for the Coast Guard's costs incurred in response to the *Deepwater Horizon* incident. Specifically, as of May 31, 2011, the Coast Guard has billed and received from responsible parties, $315.3 million for Coast Guard recoverable, or indirect costs, such as personnel and equipment. According to the agency, the Coast Guard has historically viewed its OPA recoverable costs as activities normally funded through the agency's operating expense appropriation, and thus it has not sought reimbursement for these costs from the Fund.

As shown in figure 2, the Fund has been administratively divided into two major components—the Emergency Fund and the Principal Fund—administered by the Coast Guard's NPFC. The Emergency Fund authorizes the President to make available $50 million each year to cover immediate expenses associated with mitigating the threat of an oil spill, costs of oil spill containment, countermeasures, and cleanup and disposal activities, as well as paying for other costs to initiate natural resource damage assessments. Amounts made available remain available until expended. For the *Deepwater Horizon* oil spill, the Coast Guard's Federal On-Scene Coordinator[20] used the Emergency Fund to pay for oil spill removal activities (i.e., the equipment used in removal activities and for the proper disposal of recovered oil and oil debris), and the Federal Natural Resource Damage Trustees[21] also entered into reimbursable agreements with NPFC with respect to funding for activities to initiate natural resource damage assessments. To the extent that available amounts are inadequate for an emergency (as was the case in the *Deepwater Horizon* oil spill), the Maritime Transportation Security Act of 2002 granted authority for the Coast Guard to advance up to $100 million to pay for oil spill removal activities, and that amount was advanced from the Principal Fund to the Emergency Fund.[22]

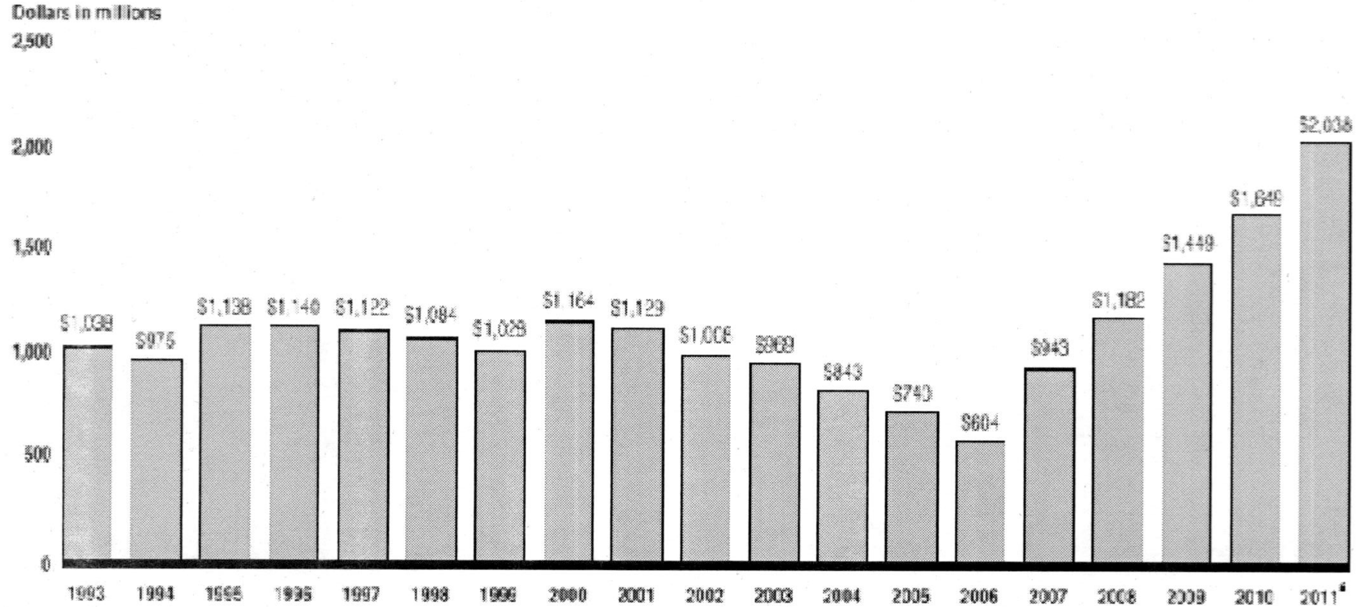

Source: GAO analysis of NPFC data.

[a]Fund balance as of May 31, 2011 (unaudited).

Figure 1. Oil Spill Liability Trust Fund Balance, September 1993–May 2011 (Unaudited).

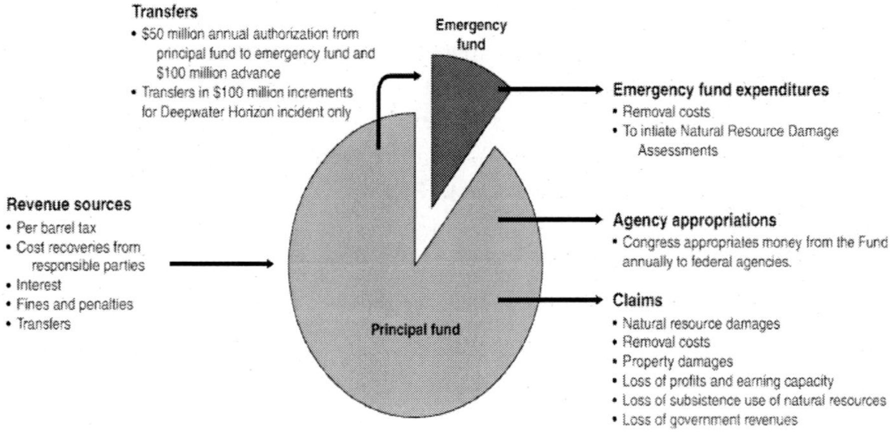

Sources: NPFC and DHS USCG Report on implementation of the Oil Pollution Act of 1990.

Figure 2. Oil Spill Liability Trust Fund Components.

In June 2010, Congress amended OPA to authorize emergency advances for the *Deepwater Horizon* oil spill in increments of up to $100 million for each cash advance, but the total amount of all advances may not exceed the $1 billion per incident cap.[23] In contrast to the Emergency Fund, the Principal Fund is to be used to provide funds for natural resource damage claims,[24] loss of profits and earning capacity claims, and loss of government revenues. The Principal Fund also provides for certain agency appropriations including the Coast Guard, Environmental Protection Agency (EPA), and the Department of the Interior (DOI)—each of which receives an annual appropriation from the Fund through the Principal Fund to cover administrative, operational, personnel, and enforcement costs.

Consistent with its Fund management responsibilities, in response to the *Deepwater Horizon* oil spill, NPFC is responsible for billing the responsible parties, including BP, directly for costs that government agencies have incurred. The payments NPFC receives from BP are to be deposited into the Fund and NPFC reimburses agencies for their removal costs.

Funds are to be disbursed from the Fund to government agencies using two vehicles—Pollution Removal Funding Authorizations (PRFA) and Military Interdepartmental Purchase Requests (MIPR). The PRFA commits the Fund to reimburse costs incurred for agreed-upon pollution response activities undertaken by a federal agency assisting the Federal On-Scene

Coordinator. The terms of a PRFA include relevant (1) personnel salary costs, (2) travel and per diem expenses, (3) charges for the use of agency-owned equipment or facilities, and (4) expenses for contractor or vendor-supplied goods or services obtained by the agency for removal assistance. Similarly, the Federal On-Scene Coordinator may issue a MIPR for agreed-upon activities of the Department of Defense (DOD) or its related components and for some other agencies' activities. In contrast to PRFAs, MIPRs generally commit the Fund to disburse funds for oil spill response activities prior to conducting the activity and incurring the related costs. However, for the *Deepwater Horizon* oil spill, both NPFC and DOD established procedures for submitting documentation on a regular basis for MIPRs authorized in response to this spill of national significance.

BP's Claims Process for Individuals and Businesses

The Coast Guard, without in any way relieving the other responsible parties it identified of liability, approved BP's advertisement of its claims process.[25] In response to economic harm caused by the *Deepwater Horizon* oil spill and to fulfill its obligations as a responsible party, BP established a claims process and multiple claims centers throughout the Gulf states. On May 3, 2010, BP began paying emergency compensation to individuals and businesses. BP stated that emergency payments would continue as long as individuals and businesses could show they were unable to earn a living because of injury to natural resources caused by the oil spill. According to BP, it would base emergency payments on 1 month of income and would be adjusted with additional documentation.[26] BP has been working to ensure that the other *Deepwater Horizon* oil spill responsible parties contribute to the response. On May 20, 2011, BP announced that it had reached an agreement with MOEX Offshore 2007 LLC and its affiliates to settle all claims between the companies related to the *Deepwater Horizon* oil spill, which included MOEX paying $1.065 billion to BP. Additionally, on October 17, 2011, BP announced that it had reached an agreement with Anadarko Petroleum Company to settle all claims between the companies related to the Deepwater Horizon oil spill, which included Anadarko paying $4 billion to BP.

On June 16, 2010, President Obama announced that BP had agreed to set aside $20 billion to pay certain economic damage claims caused by the oil spill.[27] On August 6, 2010, BP established an irrevocable Trust and committed to fund it on a quarterly basis over 3-1/2 years to reach the $20 billion total (as shown in fig. 3). The Trust is to pay some OPAcompensable claims as well as some other claims for personal injuries that are not OPA-compensable, but for which BP would be liable under other law.[28]

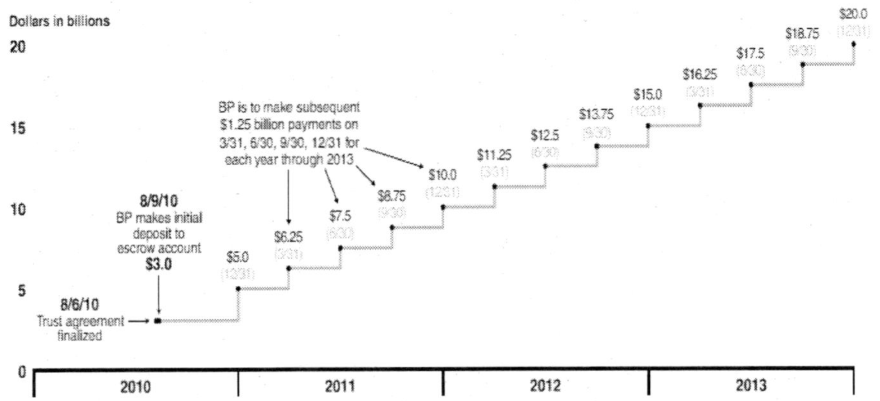

Source: GAO analysis of an August 9, 2010 BP press release on the funding of the $20 billion account.

Figure 3. The Trust's $20 Billion Funding Time Frame.

On August 23, 2010, the GCCF took over the administration of claims process and the centers BP had established.[29] Since it began operating, the GCCF has offered the following kinds of payments:

- Emergency Advance Payments: Payments available to individuals and businesses that were experiencing financial hardship resulting from damages incurred from the Deepwater Horizon oil spill. GCCF considered claims on emergency payments that were submitted by November 23, 2010.
- Quick Payment Final Claim: On December 13, 2010, BP announced that individuals and businesses that had received emergency payments from the GCCF were eligible for a quick payment final claim, which offers a fixed amount of $5,000 for individuals and $25,000 for businesses. Acceptance of such a claim would resolve all claims by that claimant against BP including past and future alleged damages. The GCCF Protocols for Interim and Final Claims provides that final claims can be submitted to the GCCF through August 23, 2013.30
- Final Payment: Those who do not choose or are not eligible for the quick payment may submit a full review final payment claim for all

documented losses and damages. Acceptance of a final claim would resolve all claims by that claimant against BP including past and future alleged damages. Under GCCF procedures, claimants will have until August 23, 2013, to estimate damages and submit claims for final payment.

- Interim Payments: The alternative to a final payment is to make an interim payment claim for past damages that have not been compensated. Individuals and businesses receiving interim payments are not required to sign a release of liability and may file a final claim at a later date. The GCCF Protocols for Interim and Final Claims provides that interim claims can be submitted to the GCCF through August 23, 2013.

As of May 31 2011, GCCF has paid $4.2 billion for individual and business claims as shown in table 1. While the GCCF is scheduled to stop receiving claims on August 23, 2013, BP's obligation, as a responsible party under OPA, to receive claims will continue after the GCCF closes.

Table 1. Claims Paid by GCCF as of May 31, 2011 (Unaudited)

Dollars in millions		
Type	**Number of claims paid**	**Amount**
Emergency Advanced Payments	169,142	$2,582.6
Interim Payments	12,977	147.8
Quick Pay (Final)[a]	114,320	1,105.4
Full Review (Final)[a]	25,102	390.5
Total[b]	**321,541**	**$4,226.3**

Source: GAO analysis of GCCF data.

[a] Both Quick Pay and Full Review require the claimants to sign a release waiving any rights they may have against responsible parties to file or participate in legal action or to submit any claim to NPFC for payment.

[b] As described in our November 2010 report, claims approved by GCCF are paid from a Trust established and funded (up to $20 billion) by BP. Prior to the establishment of GCCF, BP had received and directly paid claims from individuals and businesses totaling $396.0 million.

Although the Total Federal Financial Risk Has Not Been Determined, Actions Are Needed to Reduce Known Risks

Both the individual circumstances of the *Deepwater Horizon* incident, as well as the overall framework of how the federal government responds to oil spills, present a mix of financial risks to the Fund and the federal government. The extent of financial risks to the federal government from the *Deepwater Horizon* is closely tied to BP and the other responsible parties and guarantors. Because the federal government's Fund would pay if the responsible party (BP through its Trust, for example) did not, and given the expectation for numerous expenses to be paid from the Trust and the fact that the full amount of damages may not be fully determined for some time, the extent of any long-term financial risks for the federal government as a result of this spill is not clear. Federal agency cleanup and restoration activities are underway and agencies continue to incur costs and submit them for reimbursement. As a result, it is possible that expenditures from the Fund for Federal removal costs and claims will reach the $1 billion cap, as the cap balance was over $626 million on May 31, 2011.[31] When the cap balance reaches the total expenditure cap of $1 billion, no further payments to reimburse agencies' costs (or to pay individual or business claims if not paid by the responsible parties) can be made from the Fund, so federal agencies would no longer be able to obtain reimbursement for their costs. Finally, the federal government's longer-term ability to provide financial support in response to future oil spills is also at risk because the Fund's primary source of revenue, a tax on petroleum products, is scheduled to expire in 2017.

BP Has Committed to Paying *Deepwater Horizon* Expenses, but the Extent of the Federal Government's Financial Exposure Remains Unknown

BP has committed to set aside $20 billion to cover potential *Deepwater Horizon* oil spill expenses[32]—and has stated its intent to pay expenses over the $20 billion if needed. BP's track record for reimbursing federal agencies for their expenses to this point has been favorable. For example, as of May 31, 2011, NPFC had sent 11 invoices to all of the responsible parties covering federal and state OPA-compensable costs totaling $711 million and BP paid all 11 invoices.[33] However, until the total expenses of the *Deepwater Horizon* oil spill have been fully determined and those amounts have then been paid by and reimbursed to the federal government, the extent of any federal government financial exposure remains unknown.[34] The financial responsibility for the spill will ultimately be determined through a lengthy and complex process involving the application of different laws and regulations, and depends upon a continuation of the ability of the responsible parties to pay expenses associated with the *Deepwater Horizon* oil spill.[35]

Although BP has established a $20 billion Trust to pay claims from individuals and businesses harmed by the spill, a number of uncertainties regarding the Trust's uses may impact its ability to adequately reimburse claimants, increasing the risk that the federal government will ultimately be responsible for paying the remaining claims. Although all uncertainties—and the associated expenses—may not be known for many years, some uncertainties that are known relate to the following issues.

- The federal government has begun an extensive natural resource damage assessment process, but the associated costs have yet to be determined. In order to start the process, in May 2010, BP agreed to provide $10 million to DOI and $10 million to the National Oceanic and Atmospheric Administration (NOAA) in the Department of Commerce. Also, in April 2011, BP committed up to $1 billion from the Trust to projects to help restore damaged natural resources in the Gulf of Mexico, such as the rebuilding of costal marshes, replenishment of damaged beaches, conservation of sensitive areas for ocean habitat for injured wildlife, and restoration of barrier islands and wetlands that provide natural protection from storms. The natural resource damage assessment and restoration process will take years to complete, so the full costs for which BP and the other responsible parties are liable have yet to be determined. The National Commission on BP Deepwater Horizon Oil Spill and Offshore Drilling report estimates that fully restoring the Gulf will take $15 billion to $20 billion and over 30 years.[36] If the responsible parties are unable or unwilling to pay, then the agencies' costs for the natural resource damages, including the costs to assess and restore, rehabilitate, replace, or acquire equivalent natural resources, would need to be reimbursed from the Fund (provided that funds were still available, given the $1 billion per incident cap).
- The responsible parties also are likely to face fines and penalties which have yet to be determined and which will be levied by federal and state governments. In particular, under the Clean Water Act, liable parties face substantial administrative and civil penalties that may be imposed by EPA or DHS.[37] According to the BP Oil Spill Commission Report, the maximum Clean Water Act civil penalties could range from $4.5 billion to $21 billion.

- BP and the other responsible parties face over 500 lawsuits from the federal government, states, investors, employees, businesses, and individuals. The extent to which these lawsuits will impact the responsible parties financially is uncertain at this time since they will take years to litigate. BP has stated that it may use the Trust to pay lawsuit settlements as well as for paying claims and for natural resource damages.
- Justice is continuing to evaluate federal government costs incurred related to the *Deepwater Horizon* oil spill that are not OPA-compensable. On May 13, 2011, Justice sent the responsible parties an invoice requesting reimbursement to the federal government for $81.6 million (for agencies' costs incurred through December 2010).
-

Although BP has stated that it will pay expenses over the $20 billion, if necessary, it is uncertain how this would be accomplished over time, thus posing an element of risk to the federal government. In addition, although MOEX and its affiliates have settled with BP by paying $1.065 billion and Anadarko settled with BP which included a payment of $4 billion, other responsible parties have not reached a settlement. If BP becomes unable to pay future cleanup costs, individual and business claims, and natural resource restoration costs, the federal government may need to consider paying costs and then pursuing reimbursement from the other responsible parties.

Reaching the Fund's $1 Billion Cap Could Result in Federal Agencies Needing Additional Funding for Oil Spill Response Costs

NPFC's *Deepwater Horizon* oil spill amounts counted towards this cap was $626.1 million as of May 31, 2011, and is thereby approaching the $1 billion per-incident cap mandated by OPA. The $626.1 million consists of $128.0 million incurred by the Coast Guard and $498.1 million incurred by other agencies. Once expenditures from the Fund reach the cap, NPFC will be statutorily barred from reimbursing federal agencies for response and restoration work, or paying individuals and businesses to settle claims. Consequently, if federal agencies did not receive dedicated appropriations for oil spill costs, the federal agencies would be faced with reallocating their appropriated funding to cover oil spill costs, or seeking additional funding from Congress. In November 2010, we suggested that Congress may want to consider setting a Fund cap associated with an incident, based upon net expenditures (expenditures less reimbursements).[38]

As of May 31, 2011, government agencies continue to submit documentation of their *Deepwater Horizon* oil spill recovery costs for reimbursement from the Fund. (App. VII provides information about government agencies' authorized response costs and amounts reimbursed.) Further, although as of May 31, 2011 all individual and business claims reviewed by NPFC have been denied, claims continue to be submitted.[39] According to NPFC officials, individuals and businesses will continue to submit claims associated with the *Deepwater Horizon* oil spill for several years. In addition, the natural resources restoration process is beginning and these associated costs will accumulate over many years.

Uncertainties Regarding Future Funding Availability

Uncertainties exist regarding the primary revenue source of the Fund, which is set to expire in 2017, and the potential for future oil spills. If the Fund's primary source of revenue expires, this could affect future oil spill response and may increase risk to the federal government. Also, although the *Deepwater Horizon* oil spill was the largest oil spill disaster in U.S. history, annually over 500 spills of varying size and response occur.

- The per barrel tax revenue. A provision of The Energy Improvement and Extension Act of 2008 mandates that the Fund's primary source of revenue, a per barrel tax, is set to expire on December 31, 2017.[40] Therefore, even with substantial amounts reimbursed by BP, the Fund balance would likely decrease as a result of the expiration of its primary source of funding and the expectation of future *Deepwater Horizon* costs. This could raise the risk that the Fund would not be adequately equipped to deal with future spills, particularly one of this magnitude, and it will be important for Congress to determine a funding mechanism for the Fund going forward. The two other sources of revenue are cost recoveries from responsible parties and interest on the Fund principal from U.S. Treasury investments.[41] As we reported in September 2007,[42] the balance of the Fund generally declined from 1995 to 2006 mostly because the per barrel tax expired in December 1994 and revenue was not collected from January 1995 to March 2006.[43]

- The potential need to fund the response to future spills poses risks. The possibility of needing to respond to another spill of national significance increases the risk to the Fund and the federal government. In fiscal year 2011 alone, the Fund has already paid for 267 oil spills through May 31, 2011. According to NPFC officials, on an annual basis, approximately 500 spills with varying costs and magnitude occur. In 2007, we reported that since 1990 approximately 51 spills amounting to over $1 million have occurred, and that responsible parties and the Fund have spent between $860 million and $1.1 billion for oil spill removal costs and compensation for damages.[44] Responsible parties paid between 72 and 78 percent of these expenses, while the Fund paid the remainder. As of May 31, 2011, the Fund's balance was approximately $2.0 billion. The federal government would need to consider using other sources of funds particularly if another spill of national significance occurs and if the responsible party(ies) are unable or unwilling to pay.

THE COAST GUARD HAS EFFECTIVE CLAIMS PROCESSING AND COST REIMBURSEMENT CONTROLS, HOWEVER COULD BENEFIT BY DOCUMENTING CHANGES MADE IN CLAIMS PRACTICES

Our testing of the Coast Guard's controls over *Deepwater Horizon* claims processed as of April 30, 2011, and cost reimbursements processed as of April 20, 2011, showed that adjudicated claims processed and costs reimbursed were consistent with its procedures. The Coast Guard's operating practices in these areas have changed to reflect the largely unprecedented size and evolving scope of the *Deepwater Horizon* incident. It has updated its cost reimbursement procedures to incorporate lessons learned from the initial response to this spill and although it has not yet updated its procedures for processing claims from spills of national significance to reflect lessons learned from its experiences processing *Deepwater Horizon* claims, it has plans to do so.

Tests Show Coast Guard's NPFC Has Established Effective Controls for Processing *Deepwater Horizon* Claims and Cost Reimbursements

We found that internal controls related to the documentation, review, and adjudication of individual and business claims submitted following the *Deepwater Horizon* oil spill were operating in accordance with established policies and procedures. During the period September 1, 2010, through May 31, 2011, NPFC received 901 *Deepwater Horizon* claims totaling $238 million. Of these claims, NPFC has finalized 570, all of which resulted in a denial or a withdrawal by the claimant.

Our testing of a statistical sample of 60 out of the 432 *Deepwater Horizon* finalized claims through April 30, 2011 found that NPFC had followed its policies and procedures.[45] Specifically, all claims

- were submitted in writing, for a sum certain amount, and included the required claimant information (i.e., address, nature and extent of the impact of the incident, etc.);
- complied with OPA's order of presentment (which requires that all claims for removal costs or damages must be presented first to the responsible party for payment), and verified that claimants had filed with the responsible party first before submitting their claim to NPFC;
- included evidence submitted by the claimant, or if needed, NPFC sent a letter to the claimant requesting additional support;
- were adjudicated within the time provided by regulation;[46]
- underwent legal review and were submitted within the required time frame, if reconsideration was requested;[47] and
- when denied, were appropriately transmitted by sending a denial letter to the claimant along with a Claim Summary/Determination Form explaining the basis for denial.

However, because all finalized claims resulted in denials or withdrawals, our testing could not assess the effectiveness of NPFC's controls over payments to individuals and business claimants.

Our statistical testing of 57 of 954 *Deepwater Horizon* cost reimbursements for government oil spill response activities from the Fund between April 20, 2010, and April 20, 2011, found that in all cases NPFC had followed established policies and procedures. Specifically, NPFC

- accepted only cost reimbursement packages from government agencies with a signed PRFA or MIPR agreement in place for *Deepwater Horizon* response costs;
- determined that the Federal On-Scene Coordinator certified that all services or goods were received;
- ensured that supporting cost documentation submitted for reimbursement complied with the PRFA statement of work or MIPR agreement;
- wrote a letter to FINCEN[48] authorizing payment (signed by an NPFC Case Officer for the amount disbursed from the Fund under the appropriate PRFA or MIPR); and
- obtained supporting documentation from the government agency requesting reimbursement.

NPFC Updated Its Cost Reimbursement Procedures to Reflect Lessons Learned for *Deepwater Horizon*, and Plans to Similarly Update Its Claims Procedures

NPFC has strengthened its cost reimbursement guidance to reflect lessons learned from experiences during the initial *Deepwater Horizon* oil spill response, and officials told us they planned to take similar steps to update its claims processing guidance. Updating NPFC's claims procedures to fully reflect *Deepwater Horizon* lessons learned will be critical should another spill of national significance occur.

On April 14, 2011, NPFC issued an appendix for its cost reimbursement procedures manual modifying the procedures the agency is to follow for spills of national significance. This appendix is based on the lessons learned from addressing the unprecedented challenges posed by the *Deepwater Horizon* oil spill.[49] It provides guidance, for example, targeting some of the issues that arose related to the management of finances, including cost documentation requirements for MIPRs with DOD. Specifically, the modified procedures provide that MIPRs will be reimbursed after the cost documentation is reviewed and work completion verified.

NPFC officials told us that its current claims processing practices have also evolved since April 2010 to reflect lessons learned from the *Deepwater Horizon* oil spill. Over the past 10 years, NPFC typically received, on average, fewer than 300 claims each year. However, in light of the dramatic increase in the number of *Deepwater Horizon* oil spill claims received, NPFC refined its practices to augment its claims processing capacity. These practices included using contractors, Coast Guard reservists and, as needed, reassigning other NPFC staff. NPFC's *Standard Operating Procedures of the Claims Adjudication Division*, which have not been updated since April 2004, do not yet include specific procedures required for processing claims for a spill of national significance.[50] In particular, the procedures do not include modified practices to respond to the dramatic increase in claims filed as a result of the *Deepwater Horizon* incident. For the *Deepwater Horizon* oil spill, NPFC adopted practices involving newly developed performance indicators, past experience and continuous updates on current GCCF statistics as tools to identify the timing and extent of additional resources needed to augment its claims processing capabilities. GAO's *Standards of Internal Control in the Federal Government*[51] provide that internal control should provide for specific activities needed to help ensure management's directives are carried out.

NPFC has an opportunity to help ensure that expertise and effective practices are not lost by incorporating the lessons learned from the *Deepwater Horizon* incident in its guidance. Clearly documenting the policies and procedures used for the *Deepwater Horizon* incident would position NPFC for more effectively processing claims from any future spills of national significance by incorporating guidance, for example, on the use of performance indicators and statistics to address the size and timing of claim submissions. NPFC officials told us they are in the process of drafting an appendix for claims for spills of national significance for its individual and business claims procedures manual to document such procedures.

FEDERAL AGENCIES' OVERSIGHT EFFORTS INCLUDE MONITORING GCCF'S CLAIMS PROCESS AND PARTICIPATING IN NATURAL RESOURCE ASSESSMENTS

Soon after the *Deepwater Horizon* oil spill, the Deepwater Integrated Services Team (IST)[52] was established at the direction of the National Incident Command, under the command of the U.S. Coast Guard, and initially was responsible for monitoring BP's claims process. As Deepwater IST scaled back, its responsibilities were transitioned to relevant agencies. The oversight effort for cost reimbursement and claims activities transitioned to Justice, who continues to lead the efforts. In addition, DOI and NOAA are serving as the federal government representatives for the natural resource trustees in evaluating the environmental impact of the *Deepwater Horizon* incident.

The Role of the Deepwater Integrated Services Team Evolved during the Response Effort and Concluded in February 2011

In order to coordinate federal agencies' and departments' efforts to provide support services and initially monitor claims in response to the *Deepwater Horizon* oil spill, the IST was established with the Federal Emergency Management Agency (FEMA) leading this effort. Figure 4 shows the IST participants. IST coordinated intergovernmental efforts to monitor BP and the GCCF claims processes to promote their efficiency and effectiveness by raising awareness and ensuring accountability and positive outcomes. It also helped raise awareness of concerns related to payment policy clarity for claimants, data access and reporting, and coordination of federal and state benefits and services to avoid duplicate payments. In conjunction with the stand-down of the National Incident Command on September 30, 2010, IST began scaling back its staffing and functions and concluded the final transition of its functions to federal agencies under the agencies existing authorities and responsibilities effective February 1, 2011. For example, Justice continues to monitor the effectiveness and efficiency of the BP and GCCF claims processes, and also leads coordination efforts to connect government stakeholders with BP and GCCF as needed.

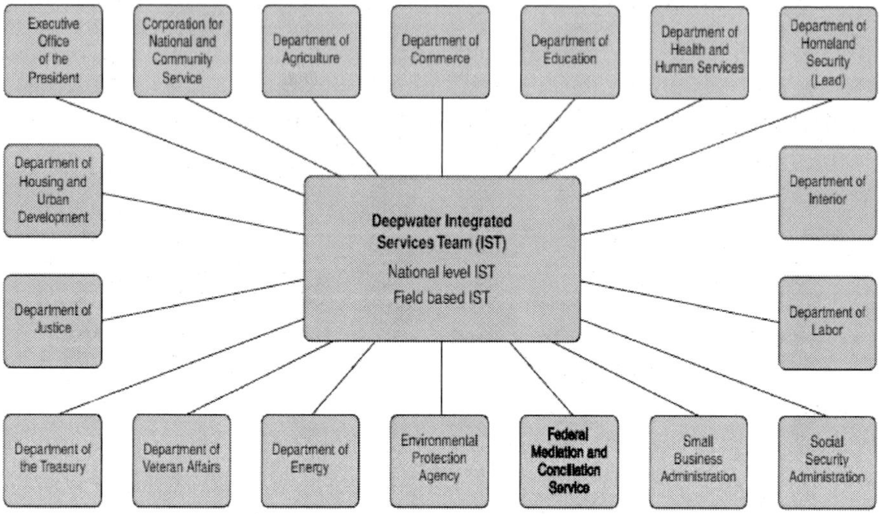

Source: Deepwater Integrated Services Team.

Figure 4. Participants in the Deepwater IST.

Justice Is Leading the Federal Government's Efforts to Monitor GCCF's Claims Process

Justice has been proactive in leading federal agencies in using a range of approaches to establish practices to monitor the cost reimbursement and claims activities of BP and the GCCF. Justice encouraged BP to establish the Trust and the GCCF. Justice sent at least four letters to GCCF highlighting key concerns with the claims process. For example, in a letter dated February 4, 2011, Justice reiterated that OPA requires BP and other responsible parties to pay for damages as a result of the oil spill and to make the GCCF claims process more transparent so that claimants clearly understand the status of their claims. According to a Justice official, Justice's involvement stems from a regulatory interest to ensure that the administration of the Trust is consistent with OPA and that claimants are treated fairly, as well as to help ensure transparency.

On another related front, in order to identify non-OPA-compensable costs which the federal government incurred due to the duration, size, and location of the *Deepwater Horizon* oil spill, OMB issued guidance between July 2010 and January 2011 to federal agencies on identifying, documenting and reporting costs associated with the *Deepwater Horizon* oil spill.[53] Specifically,

OMB's guidance directed federal agencies to include in their summary cost reports federal employee time, travel, and other related costs that were not being reimbursed through the Fund.[54] Justice has used the information submitted by the federal agencies to identify and seek reimbursement from responsible parties for certain nonOPA-compensable costs. According to Justice officials, Justice reviewed and analyzed the information submitted by the agencies through December 31, 2010, to determine which agency costs reflected agency activities directly related to the *Deepwater Horizon* oil spill. After compiling this information, on May 13, 2011, Justice sent the responsible parties an invoice requesting reimbursement to the federal government for $81.6 million for the first two reporting quarters (through approximately December 2010) for other federal agency non-OPA-compensable costs.[55] According to Justice officials, they will continue to analyze the *Deepwater Horizon* oil spill costs that federal agencies submit on a quarterly basis and plan to send additional requests for cost reimbursement to the responsible parties, as appropriate.

Justice has also coordinated investigations of *Deepwater Horizon* potential fraudulent claims from individuals and businesses under review by its National Center for Disaster Fraud.[56] As of July 28, 2011, over 3,000 referrals had been submitted for investigation from BP, GCCF and NPFC.

The National Commission on BP Deepwater Horizon Oil Spill and Offshore Drilling recommended that Justice's Office of Dispute Resolution conduct an evaluation of GCCF once all claims have been paid, in order to inform claims processes in future spills of national significance. The Commission said the evaluation should include a review of the process, the guidelines used for compensation, and the success rate for avoiding lawsuits.

NPFC has also participated in monitoring the individual and business claim activities of BP and GCCF in order to determine and prepare for any potential inflows of related claims that might be coming to NPFC following any significant number of claim denials by BP or the other responsible parties. Claimants who are denied payment by the GCCF or whose claims are not settled within 90 days may pursue the following options:

- appeal GCCF's decision, if the claim is in excess of $250,000 under procedures established by the GCCF administrator;
- begin litigation against the responsible parties in court;[57] or
- file a claim with NPFC.[58]

Over 900 *Deepwater Horizon* claims (some of which were denied by BP and GCCF) have been filed with NPFC between September 2010 and May 2011. NPFC's claims adjudication division regularly obtains information from GCCF on GCCF claims paid and denied. This oversight information allows NPFC to determine the extent to which cases should be closed as the claimants were paid by GCCF, helps prevent claimants being paid by both GCCF and NPFC for the same claim, and enables it to better anticipate denied GCCF claims that could be resubmitted to NPFC for adjudication.

DOI and NOAA Are Actively Participating as the Federal Natural Resource Trustees

The natural resource trustees for the *Deepwater Horizon* incident—responsible for evaluating the oil spill's impacts on natural resources—are DOI, NOAA, DOD, and the five Gulf Coast states (Texas, Louisiana, Mississippi, Alabama, and Florida). On September 27, 2010, NOAA sent the eight responsible parties identified by DOI a Notice of Intent to Conduct Restoration Planning for the *Deepwater Horizon* incident on behalf of federal and state trustees. On April 21, 2011, the federal and state trustees announced that BP had agreed to provide \$1 billion from the Trust for early restoration projects in the Gulf of Mexico to address natural resource damage caused by the *Deepwater Horizon* oil spill. Under the agreement, the \$1 billion will be provided to fund projects such as the rebuilding of coastal marshes, replenishment of damaged beaches, conservation of sensitive areas for ocean habitat for injured wildlife, and restoration of barrier islands and wetlands that provide natural protection from storms.

The \$1 billion in early restoration projects will be selected and implemented as follows:

- DOI will select and implement \$100 million in projects;
- NOAA will select and implement \$100 million in projects;
- each of the five states (Alabama, Florida, Louisiana, Mississippi, and Texas) will select and implement \$100 million in projects; and
- DOI and NOAA will select projects submitted by the state trustees for \$300 million.

Conclusion

Several factors contribute to financial risks that the federal government will continue to face for a number of years as a result of the *Deepwater Horizon* oil spill. Future uncertainties include the total expenses of fully addressing the impact of the *Deepwater Horizon* oil spill and the responsible parties' and guarantors' willingness and ability to continue to pay, possibly for the next several decades. Uncertainty over federal financial risks also arise from the per barrel oil tax expiration in 2017—the primary revenue source for the Fund—and the need for funding in response to other potential significant spills. Given these risks, it will be important for Congress to consider whether additional legislative action would help ensure that OPA's $1 billion per-incident cap does not hinder NPFC's ability to reimburse federal agencies' costs, pay natural resources damages, and pay valid claims submitted by individuals and businesses. To this end, we are reiterating the Matter for Congressional Consideration in our November 2010 report that Congress should consider amending OPA, or enacting new legislation to take into account reimbursements from responsible parties in calculating an incident's expenditures against the Fund's $1 billion per-incident expenditure cap.[59] For its part, NPFC has an opportunity to document and incorporate the lessons learned from its *Deepwater Horizon* oil spill experience in its policies and procedures to help improve its management of any future spills of national significance. Capturing lessons learned about processing such claims will be essential should a significant spill occur in the future In addition, NPFC took action to address recommendations made in our November 2010 report to ensure and maintain cost reimbursement policies and procedures and ensure responsible parties are properly notified (see app. I for the recommendations and their current status).

Matter for Congressional Consideration

Congress should consider the options for funding the Oil Spill Liability Trust Fund as well as the optimal level of funding to be maintained in the Fund, in light of the expiration of the Fund's per barrel tax funding source in 2017.

Recommendation for Executive Action

In order to provide guidance for responding to a spill of national significance and build on lessons learned, we recommend that the Secretary of Homeland Security direct the Director of the Coast Guard's NPFC to finalize the revisions the Coast Guard is drafting to its *Claims Adjudication Division's Standard Operating Procedures* to include specific required steps for processing claims received in the event of a spill of national significance.

Agency Comments

We provided copies of the draft report to the Departments of Homeland Security, Justice, Interior, Defense, and Commerce; Office of Management and Budget; and Environmental Protection Agency for comment prior to finalizing the report. In its written comments, the Department of Homeland Security concurred with our recommendation and stated it plans to finalize changes to operating procedures by October 31, 2011. The Departments of Homeland Security, Justice, and Interior and Environmental Protection Agency also provided technical comments that were incorporated, as appropriate.

Susan Ragland
Director
Financial Management and Assurance

List of Requesters

The Honorable John Conyers, Jr.
Ranking Member
Committee on the Judiciary
House of Representatives

The Honorable Bennie G. Thompson
Ranking Member
Committee on Homeland Security
House of Representatives

The Honorable Tom Carper
Chairman
Subcommittee on Federal Financial Management,
Government Information,
Federal Services, and International Security
Committee on Homeland Security
and Governmental Affairs United States Senate

The Honorable Sheldon Whitehouse
Chairman
Subcommittee on Oversight
Committee on Environment and
Public Works United States Senate

The Honorable Cliff Stearns
Chairman
Subcommittee on Oversight and Investigations
Committee on Energy and Commerce
House of Representatives

The Honorable Mary Landrieu
United States Senate

The Honorable Michael C. Burgess

The Honorable Nick J. Rahall,
II House of Representatives

APPENDIX I: STATUS OF PRIOR RECOMMENDATIONS

The National Pollution Fund Center (NPFC) took actions as of September 2011 to address the four recommendations we made in our November 2010 report.[1]

Table 2. Implementation Status of Prior GAO Recommendations

GAO recommendations	Implementation status
In order to help establish and maintain effective cost reimbursement policies and procedures for the Fund, we recommended that the Secretary of Homeland Security direct the Director of the Coast Guard's NPFC to update NPFC's policies and procedures to include:	
1.Current Fund reimbursement-billing practices that reflect both a percentage of federal agencies' obligations as well as expenditures.	NPFC officials stated they updated their policies and procedures to formally incorporate the practices in April 2011.
2.Specific procedural guidance on processing Department of Defense (DOD) requests for reimbursement using Military Interdepartmental Purchase Requests.	NPFC officials stated they updated their policies and procedures to formally incorporate the procedures in April 2011.
In order to ensure that responsible parties are properly notified of their responsibilities for an oil spill, we recommend that the Secretary of Homeland Security direct the Director of NPFC to:	
3.Update NPFC's current policies to reflect current organization and structure and management's directives.	NPFC officials stated that this recommendation resulted from outdated procedures regarding Notices of Designation and the procedures were corrected in August 2011.
4.Update NPFC's current procedures to provide detailed guidance and procedures for identifying and documenting responsible party notification.	In their comments to GAO's November 2010 report, NPFC officials disagreed with our recommendation and stated its responsible party designations are unrelated to the imposition of liability under the Oil Pollution Act of 1990 (OPA) and that they serve the purpose of getting a responsible party to advertise the Deepwater Horizonoil spill claims process. NPFC's procedures provided that responsible parties and their guarantors are to be notified of their oil spill-related responsibilities. In accordance with its procedures, NPFC sent formal letters of designation to some, but not all, of the responsible parties it identified for the Deepwater Horizonoil spill. To other responsible parties, NPFC provided only invoices that reflected NPFC's assessment of liability for removal costs. NPFC's procedures for notifying responsible parties using invoices did not clearly communicate their "responsible party" designation. NPFC officials stated that NPFC updated its procedures in August 2011 to clarify the process.

Source: GAO-11-397R and GAO analysis of NPFC information.

APPENDIX II: OBJECTIVES, SCOPE, AND METHODOLOGY

This report is the third and final in a series of reports on the *Deepwater Horizon* oil spill in response to this request. Shortly after the explosion and subsequent sinking of BP's leased *Deepwater Horizon* oil rig in the Gulf of Mexico in April 2010, we were requested to (1) identify the financial risks to the federal government and, more specifically, to the Oil Spill Liability Trust Fund (Fund) resulting from oil spills, particularly *Deepwater Horizon*, (2) assess NPFC's internal controls for ensuring that processes and payments for cost reimbursements and processes for claims related to the *Deepwater Horizon* oil spill were appropriate, and (3) describe the extent to which the federal government oversees the BP and Gulf Coast Claims Facility (GCCF) *Deepwater Horizon* oil spill-related claims processes.

Concerning our analysis of the financial risks and exposures to the federal government and Fund, we identified and analyzed applicable laws and regulations in order to determine statutory and regulatory limitations on the liability of responsible parties that may pose financial risks to the Fund and federal government. We also considered GAO reports on the use of the Fund, reviewed publicly available quarterly financial information of responsible parties through June 2011 to gain an understanding of the extent to which contingent liabilities are reported by these companies, and reviewed reports issued by the Congressional Research Service on responsible party liabilities under OPA. [1] To determine the obligations and costs incurred in relation to the Fund's $1 billion per incident cap, we obtained and analyzed daily financial summary data NPFC used related to the *Deepwater Horizon* oil spill. We also reviewed NPFC's daily financial summary data to compare the amounts federal and state agencies had submitted for reimbursement from the Fund to the amounts NPFC had authorized for payment from the Fund to these government agencies through May 2011. We obtained invoices NPFC sent to the responsible parties to reimburse the Fund, analyzed the requests for reimbursements submitted by federal and state agencies, and compared the invoiced amounts to the amounts federal and state agencies had submitted for payment from the Fund.

To assess NPFC's internal controls for ensuring that agencies' requests for cost reimbursements and claims from individuals and businesses are appropriate, we reviewed relevant sections of OPA and compared the sections to NPFC's cost reimbursement and claims Standard Operating Procedures and to GAO's *Standards for Internal Control in the Federal Government.*[2] We interviewed cognizant NPFC officials about its cost reimbursement and claims processes, *Deepwater Horizon* oil spill response efforts, specific cost recovery actions under way or completed, and the NPFC division(s) responsible for those actions. We also conducted walkthroughs of the cost reimbursement and claims processes, observed NPFC's process for generating an invoice to the responsible parties for *Deepwater Horizon* response costs, and conducted a site visit to the Gulf area in October 2010.

For agency cost reimbursements, we tested a statistical sample of payments to federal and state agencies for their *Deepwater Horizon* removal and response activities paid from the Fund between April 2010 and April 2011. We interviewed NPFC's Case Management Officer for *Deepwater Horizon* and other NPFC officials to gain a thorough understanding of NPFC's cost reimbursement process. In addition, we performed walk-throughs of NPFC's cost reimbursement and billing processes and reviewed NPFC's Case Management's standard operating procedures and other guidance documents. We also obtained updated information from NPFC officials about the status of the response to recommendations made in our November 2010 report.

To determine our population for sampling cost reimbursements for the *Deepwater Horizon* oil spill, we obtained a disbursement file from U.S. Coast Guard's Finance Center (FINCEN)[3] which consisted of 173,458 disbursements from the Fund between April 2010 and April 2011. We reviewed the information in the file to determine whether we could rely on the data in order to select a sample and test internal controls associated with the cost reimbursement process. We assessed the reliability of the data in the file and determined it could be used to select a statistical sample for testing. From the population of 173,458 disbursements from the Fund between April 2010 and April 2011, we identified 954 disbursements for *Deepwater Horizon*.[4] We then selected a random statistical sample of 57 disbursements for testing. We tested the 57 Fund disbursements for adherence to NPFC's case management standard operating procedures. Our test included reviewing the request for reimbursement submission to

- determine if a signed Pollution Removal Funding Authorization (PRFA) or Military Interdepartmental Purchase Request (MIPR) was in place between the performing federal or state agency and the Federal On-Scene Coordinator;
- assess that the services or goods provided were in accordance with the terms of the PRFA statement of work or MIPR agreement;
- confirm evidence of supporting documentation;

- confirm the Federal On-Scene Coordinator's approval of the amount requested for reimbursement by the performing federal or state agency; and
- confirm an NPFC Case Manager signed an Authorization to Pay or Authority to Allow Intra-Governmental Payment and Collection memorandum addressed to FINCEN authorizing payment from the Fund.

For claims, we tested a statistical sample of finalized *Deepwater Horizon* claims presented to the Fund between September 2010 and April 2011. First, we interviewed NPFC's Claims Division Chief, Senior Claims Manager, and other cognizant NPFC officials to gain an understanding of NPFC's claims adjudication process. On the basis of information provided by NPFC, we identified 432 finalized claims from NPFC's Claims Processing System[5] submitted for the *Deepwater Horizon* spill between September 2010 and April 2011. From the population of 432 finalized claims, we selected a random sample of 60 claims to test. We tested the sample for adherence to OPA's and NPFC's claims policies and procedures. We tested NPFC's adherence to its procedures for claim receipt, initial review, adjudication review, determination, and reconsideration. In conducting our work, we reviewed documents from individual claim files, and also used NPFC's Claims Processing System to review the responsible party's communication on the claims presented to the NPFC for payment. We tested to ensure that NPFC had a process for complying with OPA's prioritization requirement that all claims be presented to the responsible party before they can be presented to the Fund. We tested to confirm that the claims were signed and submitted in writing, for a sum certain amount, and were processed by NPFC within the required statutory time frame. Because there were no payments made for claims submitted for *Deepwater Horizon* for our scope period, we were unable to test the payment process.

Because we selected a sample of claims and cost disbursements, our results are estimates of the population and thus are subject to sample errors that are associated with samples of this size and type. Our confidence in the precision of the results from these samples is expressed in 95-percent confidence intervals. A 95-percent confidence interval is the interval that would contain the true population value in 95 percent of samples of this type and size. The results of our tests on both the sample of claims and the sample cost disbursements did not find any exceptions. On the basis of these results, we estimated that the 95-percent confidence intervals range from zero to 5 percent for both sample results and concluded with 95-percent confidence that the error rate in each population does not exceed 5 percent.

We reviewed NPFC's policies and procedures for processing and adjudicating oil spill claims and obtained information on NPFC's claims contingency planning for handling potential surges in claims submitted related to the *Deepwater Horizon* oil spill.

We obtained claims information from the GCCF and NPFC through May 2011 to describe the number and types of claims filed by individuals and businesses against the GCCF and the Fund, and the number and dollar amounts submitted, reviewed, and paid. We also obtained the Notices of Designation[6] NPFC sent to responsible parties and interviewed NPFC officials about their methodology for identifying responsible parties and their procedures for notifying them.

We interviewed officials at the Departments of Commerce, Defense, Interior, and Homeland Security, and the Environmental Protection Agency to obtain an understanding of these agencies' response activities for the *Deepwater Horizon* oil spill and its process for billing on costs incurred. We also obtained invoices NPFC sent to the responsible parties and analyzed these billed amounts and summarized the amounts by federal and state agencies. We compared the amounts submitted for reimbursement from the Fund by the performing federal and state agencies, to the amounts billed to the responsible parties on their behalf to identify which agencies have begun their cost recovery efforts. We compared the amounts requested for reimbursement from the Fund by the performing federal and state agencies, to the amounts reimbursed from the Fund to determine the status of agency's cost recovery efforts.

To describe how the federal government oversees the BP and GCCF claims processes, we interviewed Department of Justice (Justice) officials about their oversight of BP's claims process, the establishment of BP's $20 billion Trust, and the setup of the GCCF. We reviewed Justice's comments on the draft GCCF Emergency Advanced Payment and GCCF Final Payment protocols, and we obtained and reviewed the Trust agreement. We obtained and reviewed letters sent by Justice to the responsible parties discussing their financial responsibilities in connection with the *Deepwater Horizon* oil spill, which requested that the responsible parties provide advance notice of any significant corporate actions related to organization, structure, and financial position. We obtained and reviewed letters sent by Justice to the GCCF highlighting concerns about its pace for processing claims, need for transparency, and compliance with OPA standards. In addition, we interviewed Deepwater Integrated Services Team (IST) officials about their coordination activities regarding the BP and GCCF claims process and social services coordination efforts. The IST which was established in June 2010 and stood down in September 2010, took steps to raise awareness of concerns related to claim payment policy clarity, data access and reporting of overall claims information, and the coordination of federal/state benefits and services to avoid duplicate payments. We reviewed documentation from the Deepwater IST including its coordination plan, team updates, and transition plan. We did not evaluate the effectiveness of the monitoring and oversight efforts by Justice and the Deepwater IST. Furthermore, we reviewed publicly available claim reports from BP and GCCF for claim amounts paid, but we did not test the claims data or amounts reported by BP or GCCF. We also interviewed Office of Management and Budget and Justice officials about their role and planned actions in collecting and reviewing agency quarterly cost submissions to bill the responsible parties on behalf of the federal government.[7]

We conducted this performance audit from July 2010 to October 2011 in accordance with generally accepted government auditing standards. Those standards require that we plan and perform the audit to obtain sufficient, appropriate evidence to provide a reasonable basis for our findings and conclusions based on our audit objectives. We believe that the evidence obtained provides a reasonable basis for our findings and conclusions based on our audit objectives.

Appendix III: National Pollution Fund Center's Individual and Business Claims Process

Overview

OPA provides for the payment of claims for uncompensated removal costs and certain damages caused by the discharge, or substantial threat of discharge, of oil into or upon the navigable waters of the U.S., its adjoining shorelines, or the Exclusive Economic Zone of the U.S. Adjudication and payment of claims for certain uncompensated removal costs and damages are paid out of the Principal Fund of the Fund.

Order of Presentment and Time Limitation for Submitting Claims to NPFC

Claims for removal or damages may be presented first to the Fund only in the following situations: NPFC has advertised or notified claimants in writing; by a responsible party who may assert a claim; by a governor of a state for removal costs incurred by the state;[1] and by a U.S. claimant in a case where a foreign offshore unit has discharged oil causing damage for which the Fund is liable.

In all other cases where the source of the discharge can be identified, the claimant must first present their OPA claim to the responsible party for payment. If the responsible party denies the claim the claimant may submit the claim to NPFC for adjudication. Regardless of specific action to deny the claim, if the responsible party is unable or unwilling to pay the claim within 90 days of the claimant's submission, the claimant may then submit the claim to NPFC for adjudication. If the responsible party denies a claim that is subsequently processed and payment is made from the Fund, NPFC will seek to recover these costs from the responsible party. Damage claims must be made within 3 years of when the damage and its connection to the spill were reasonably discoverable with the exercise of due care. Claims for removal costs must be made within 6 years after the date of completion of all removal actions for the incident.[2]

Designation of the Source of the Incident, Responsible Party Notification, and Advertisement

The process of designating the source of an oil discharge and notifying the responsible party frequently advances concurrently with the Federal On-Scene Coordinator's attempt to identify the responsible party during the initial stages of spill response. In addition to the Federal On-Scene Coordinator issuing a letter of Federal Interest,[3] the Federal On-Scene Coordinator and NPFC's Case Management and Claims Division[4] may decide that the potential for claims exists. Once decided, the Claim Manager is normally responsible for executing the Notice of Designation.[5] Designation of a responsible party may also occur immediately following an on-site visit or more incrementally as information on the identity of the responsible party becomes available.

Claimant Requirements

While NPFC has a form which claimants may use to submit their claim, there is no required format for submitting a claim to NPFC. However, OPA through its implementing regulations, requires that the claim be (1) submitted in writing, (2) for a sum certain amount of compensation for each category of uncompensated damages or removal, and (3) signed by the claimant. The claimant bears the burden of providing all evidence, information, and documentation deemed necessary by NPFC to support the claim. While the claim is pending against the Fund, if the claimant receives any compensation for the claimed amounts, the claimant is required to immediately amend the claim submitted to NPFC.

Table 3. List of OPA-Compensable Claim Types/Description, Eligibility, and NPFC Claims Division Responsible for Processing the Claim

Claim type	Description	Eligible claimant	Responsible NPFC claims division
Natural Resource Damages	Damages for injury to, destruction of, or loss of natural resources, including the reasonable costs of assessing the damage.	Federal, state, foreign and Indian tribal trustees.	Natural ResourceDamage Claims Division
Real or Personal Property	Damages or economic loss related to the destruction or harm of real or personal property presented by either a claimant owning or leasing the property. Does not include personal injury.	Person or entity who owns or leases property.	Claims Adjudication Division
Removal Costs[a]	Costs to prevent, minimize, or mitigate oil pollution.	Anyone incurring removal costs.	Claims Adjudication Division
Loss of Subsistence Use of Natural Resources	Damages resulting from the injury, destruction, or loss of natural resources used by the claimant to obtain food, shelter, clothing, medicine, or other minimum necessities of life.[b]	Claimant who actually uses for subsistence, the natural resources which have been injured, destroyed, or lost, without regard to the ownership or management of the resources.	Natural ResourceDamage Claims Division
Loss of Government Revenues	Net loss of taxes, royalties, rents, fees, or net profit shares due to the injury, destruction, or loss of real property, personal property, or natural resources.	Federal government, state, or a political subdivision of a state.	Claims Adjudication Division

Table 3. (Continued)

Claim type	Description	Eligible claimant	Responsible NPFC claims division
Loss of Profits and Earning Capacity	Damages equal to the loss of profits or impairment of earning capacity due to the injury, destruction, or loss of property or natural resources.	Claimant sustaining the loss or impairment.	Claims Adjudication Division
Cost of Increased Public Services	Net costs of providing increased or additional public services during or after removal activities, including protection from fire, safety, or health hazards caused by a discharge of oil.	State or political subdivsion of a state.	Claims Adjudication Division
Claims by a Responsible Party	Claims submitted by a responsible party are not processed like other OPA claims. A responsible party may present a complete defense or limitation of liability claim to NPFC for removal costs and damages paid under the provisions of OPA. Claims that meet the initial review and preliminary screening must first be evaluated to determine "entitlement" to a complete defense or limit of liability. Once entitlement has been granted, the underlying cost portion of the claim may be measured and adjudicated.	Responsible party who establishes entitlement to a defense to liability or limitation of liability in accordance with OPA (33 U.S.C. 2703-04 and 2708).	Claims Adjudication Division

Source: GAO analysis of NPFC information.

[a] Claimant must establish that the actions taken were necessary, removal costs were incurred as a result of these actions, and the actions taken were determined by the Federal On-Scene Coordinator to be consistent with the National Contingency Plan or were directed by the Federal On-Scene Coordinator. This is the most common claim type received by NPFC.

[b] Compensation allowable is based on the reasonable replacement cost of the natural resource needed during the loss period for subsistence, less all compensation made available for subsistence loss, all income received by using the time otherwise for subsistence, and all overhead or other normal expenses for subsistence use that was avoided as a result of the incident.

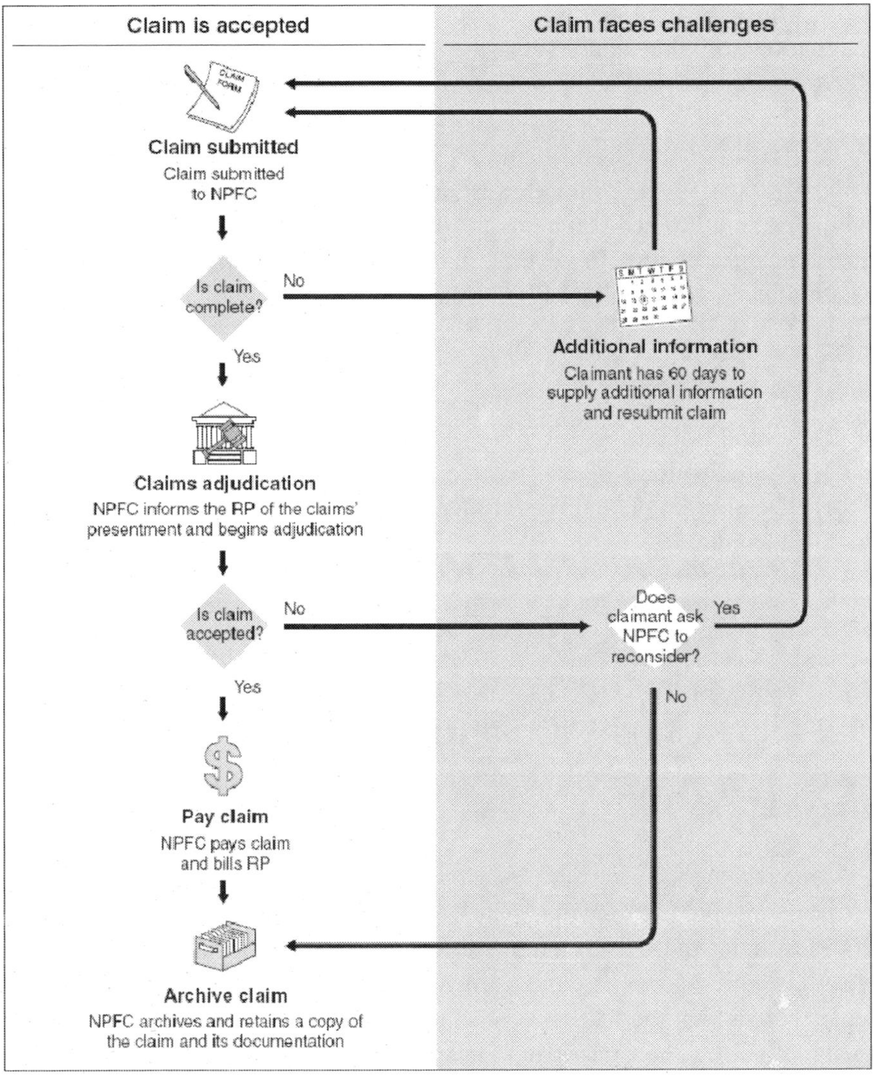

Source: GAO analysis of NPFC's Claims Adjudication Process.

Figure 5. NPFC's Claims Adjudication Process.

NPFC's Claims Process

NPFC has established standard operating procedures[6] for the activities its Claims Division undertakes throughout its process for receiving and adjudicating claims. For some oil spill incidents, the Claims Division activities begin prior to the submission of any claims. These activities include designation of the source of the spill, responsible party notification, advertisement, as well as a number of on-site activities. As noted in the table above, the Claims Adjudication Division accepts claims for uncompensated removal costs incurred and damages suffered as a result of an oil pollution incident, whereas the Natural Resource Damage Claims Division[7] accepts claims from authorized claimants for damages to natural resources and loss of subsistence use claims. In general, regardless of which division is responsible for adjudicating the claim, NPFC follows the same steps in processing these claims. [8]

1) Claim Receipt and Assignment
2) Initial Review
3) Adjudication Review
4) Determination and Reconsideration
5) Payment
6) Archive

APPENDIX IV: NATIONAL POLLUTION FUND CENTER'S COST REIMBURSEMENT PROCESS

Overview

Among other duties, the U.S. Coast Guard's NPFC administers the Fund by disbursing funds to federal, state, local, or tribal agencies for their removal activities under the Oil Pollution Act of 1990, as amended (OPA).[1] When an oil spill occurs, relevant federal agencies are notified by the National Response Center[2] including the U.S. Coast Guard and the Environmental Protection Agency (EPA).[3] The Coast Guard has responsibility and serves as the Federal On-Scene Coordinator for spills occurring in the coastal zones, while EPA has responsibility for spills that occur on land.

NPFC's Case Management Division is responsible for providing access to the Emergency Fund when a spill occurs and for working with the Federal On-Scene Coordinator and agencies to ensure accurate cost documentation to support cost recovery. NPFC's Case Management Division operates through a matrix organization comprised of four regional case teams. Each regional case team consists of a manager and multiple case officers. When a spill occurs, NPFC assigns responsibility to the regional case team representing the geographic area in which the spill occurs.

NPFC uses a three-level system to help determine the complexity of an oil spill case and its required documentation for cost reimbursement. Level I (Routine) represents about 85 percent of all oil spill incidents, in which total removal costs to the government are not expected to exceed $50,000, removal activities are localized, and removal activities can be completed within 2 weeks. For a Level I incident, agencies submit documentation to the Federal On-Scene Coordinator at the completion of removal activities. Level II (Moderately Complex) represents about 10 to 15 percent of all oil spill incidents, in which total removal costs to the government are not expected to exceed $200,000. Level II removal activities take place in multiple locations, require the involvement of several external resources (i.e., state agencies and other government units), and removal activities take longer than 2 weeks to complete. Level III (Significantly Complex) represents less than 5 percent of all oil spill incidents with total removal costs greater than $200,000. Level III removal activities take place in multiple locations, require the involvement of numerous contractors, and similar to Level II, the assistance of several external resources is needed. For both Level II and III incidents, documentation is submitted to the Federal On-Scene Coordinator as often as practical (daily if possible) until final removal activities are completed. Because the Federal On-Scene Coordinator is considered the best judge of factors regarding the oil spill, he or she is expected to select the level of documentation appropriate for the situation.

The Federal On-Scene Coordinator is responsible for issuing PRFAs or MIPRs to obtain removal and logistical services from other government agencies.

The PRFA commits the Fund to payment, by reimbursement, of costs incurred for agreed-upon pollution response activities undertaken by a federal agency assisting the Federal On-Scene Coordinator. The terms of a PRFA may include (1) salary costs, (2) travel and per diem expenses, (3) charges for the utilization of agency-owned equipment or facilities, and (4) expenses for contractor- or vendor-supplied goods or services obtained by the agency for removal assistance. Similarly, the Federal On-Scene Coordinator may issue a MIPR for agreed-upon activities of the DOD or its related components. In contrast to PRFAs, MIPRs (used primarily by DOD and its components)[4] commit the Fund to reimburse costs based on valid obligations incurred for oil spill response activities prior to being incurred. For the *Deepwater Horizon* oil spill, NPFC's cost reimbursement documentation requirements are the same for both MIPRs and PRFAs. Differences between PRFAs and MIPRs include that PRFAs are a reimbursement agreement and require the agency to submit documentation demonstrating services and have the Federal On-Scene Coordinator certify completion of work, prior to NPFC disbursing funds to the agency. For other than *Deepwater Horizon*, MIPRs allow DOD to receive the funds from NPFC prior to submitting documentation or obtaining certification of completion of work.

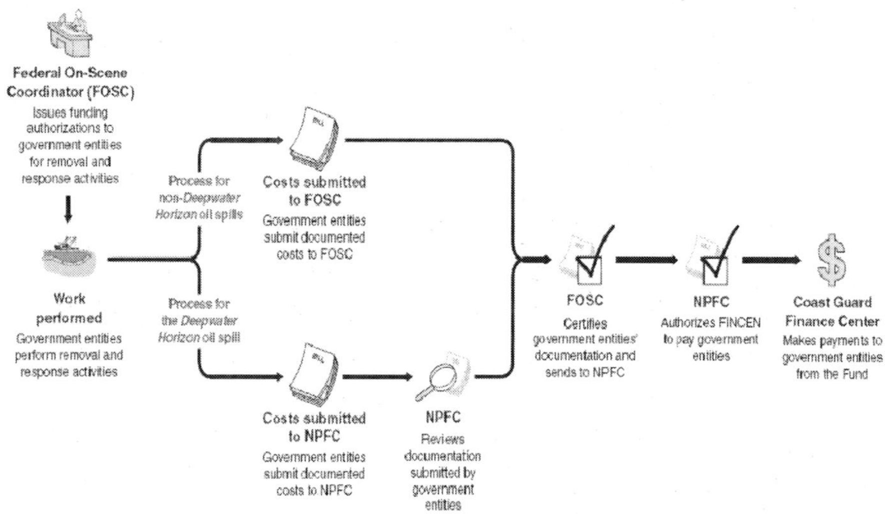

Source: GAO analysis of Cost Reimbursement Process.

Figure 6. NPFC's Cost Reimbursement Process.

The following are the six major steps for NPFC's cost reimbursement process for federal, state, and local government agencies requesting payment from the Fund.

1) Federal On-Scene Coordinator issues PRFA or MIPR to government agency.
2) Government agency performs oil spill removal and response activities and submits reimbursement request to the Federal On-Scene Coordinator.
3) Federal On-Scene Coordinator reviews and certifies that services were provided by the government agency.

4) Federal On-Scene Coordinator forwards agency's reimbursement request to NPFC for review and approval.
5) NPFC reviews agency's reimbursement documentation and sends Authorization-to-Pay memorandum to FINCEN[5] approving payment from the Fund.
6) FINCEN reimburses government agency for its oil spill removal costs.

APPENDIX V: TYPES OF OIL POLLUTION ACT-COMPENSABLE REMOVAL COSTS AND DAMAGES

Removal costs	
Removal of oil	Costs for the containment and removal of oil from water and shorelines including contract services (such as cleanup contractors and incident management support) and the equipment used for removal.
Disposal	Costs for the proper disposal of recovered oil and oily debris.
Personnel	Costs for government personnel and temporary government employees hired for the duration of the spill response, including costs for monitoring the activities of the responsible parties.
Prevention	Costs for the prevention or minimization of a substantial threat of an oil spill.
Damages	
Natural resources	Federal, state, foreign, or Indian tribe trustees can claim damagesfor injury to, or destruction of, and loss of, or loss of use of, natural resources, including the reasonable costs of assessing the damage.
Real or personal property	Damages for injury to, or economic loses resulting from destruction of, real or personal property.
Subsistence use	Damages for loss of subsistence use of natural resources, without regard to the ownership or management of the resources.
Government revenues, profits, and earning capacity	The federal, state, or local government can claim damages for theloss of taxes, royalties, rents, fees, or profits. Companies can claim damages for loss of profits or impairment of earning capacity.
Public services	States and local governments can recover costs for providing increased public services during or after an oil spill response, including protection from fire, safety, or heath hazards.

Source: GAO summary of the Oil Pollution Act of 1990 (33 U.S.C. 2702 (b)).

APPENDIX VI: INSPECTORS GENERAL ARE REVIEWING AGENCIES' *DEEPWATER HORIZON* OIL

Spill Costs

DHS, EPA, and the Department of Commerce inspectors general (IG) performed or are performing work related to their agency's costs to respond to the *Deepwater Horizon* oil spill. The DHS IG is performing an audit to determine whether the Coast Guard has adequate policies, procedures, and controls in place to capture all direct and indirect costs associated with the *Deepwater Horizon* oil spill. The EPA IG is conducting work to determine if EPA has adequate controls in place to recover its Gulf Coast oil spill response costs.

The Department of Commerce IG has published a review of the National Oceanic and Atmospheric Administration's (NOAA) tracking of oil spill costs.[1] In December 2010, the Department of Commerce IG found that while NOAA had developed processes to track the costs associated with its *Deepwater Horizon* oil spill activities, improvements are needed to ensure that all costs charged to oil spill projects—whether funded by appropriations or reimbursements—are properly recorded in the financial system and supported by sufficient, appropriate documentation. NOAA's official comments emphasized the unprecedented mobilization as a result of the scope of the *Deepwater Horizon* oil spill, and stated that as NOAA's participation has become more routine, its documentation of the oil spill activities has become more consistent. In addition, as NOAA evaluates its own execution of the response process, NOAA stated it will examine the observations provided by the IG.

APPENDIX VII: AGENCIES AUTHORIZED AND REIMBURSED COSTS FOR *DEEPWATER HORIZON* OIL SPILL RESPONSE EFFORTS

To determine the extent to which government agencies have been reimbursed from the Fund for their *Deepwater Horizon* response efforts, we obtained and analyzed reimbursement information from NPFC from April 2010 through May 2011. We found that the total maximum amount authorized through intergovernmental agency agreements for federal agencies' and states'

Deepwater Horizon oil spill response costs is over $477.7 million. However, only seven federal agencies[1] have submitted and received payment from the Fund totaling $189.4 million for their response costs; and six federal agencies[2] that have an agreement in place authorizing them to perform work and receive reimbursement from the Fund for their response efforts, have either not yet submitted a request for reimbursement or have not provided sufficient supporting documentation for their request. (See table 4.)

Table 4. Agencies' *Deepwater Horizon* Authorized Response Costs and Reimbursements Received as of May 31, 2011

Agency	Amount authorized (ceiling amount)	Amount reimbursed (actual expenditures)	Amount reimbursed to amount authorized
Department of Defense	$163,700,489	$98,125,651	60%
Department of Interior	93,367,928	11,743,756	13%
Department of Commerce	76,962,059	25,437,859	33%
Environmental Protection Agency	61,920,863	36,002,465	58%
States[a]	40,320,984	29,854,662	74%
Department of Homeland Security	16,998,513	11,949,732	70%
Department of Energy	9,056,712	4,301,033	47%
Department of Health and Human Services	8,849,859	-	0%
Department of Labor	3,260,663	-	0%
Department of Agriculture	3,083,929	1,862,479	60%
Department of Justice	141,680	-	0%
National Transportation Safety Board	24,640	-	0%
National Security Agency	18,480	-	0%
Office of the Director of National Intelligence	12,320	-	0%
Total	$477,719,119	$219,277,637	46%

Source: GAO analysis of NPFC-provided information.

[a]The amount shown for states does include amounts for certain localities that represent less than 1 percent of the total.

End Notes

[1] BP America Production Company, a subsidiary of BP p.l.c., leased the *Deepwater Horizon* from Transocean Holdings LLC, a subsidiary of Transocean Limited. Transocean Limited is the world's largest offshore drilling contractor comprising numerous subsidiaries and jointly controlled entities and associates. Unless otherwise referring to specific subsidiaries or affiliates, we refer to Transocean Limited and its components separately or jointly as "Transocean." BP p.l.c. is an international oil and gas company comprising numerous subsidiaries and jointly controlled entities and associates. Unless otherwise referring to specific subsidiaries or affiliates, we refer to BP p.l.c. and its components separately or jointly as "BP." BP was originally incorporated in 1909 in England and Wales as "British Petroleum" and changed its name in 2001.

[2] Pub. L. No. 101-380, 104 Stat. 484 (Aug. 18, 1990).The Coast Guard identified the following companies as responsible parties or guarantors for the *Deepwater Horizon* oil spill: BP Exploration & Production, Inc.; BP Corporation North America, Inc.; Anardarko, E&P Company, LP; Anardarko Petroleum Corporation; MOEX Offshore 2007 LLC; Transocean Holdings Incorporated; and QBE Underwriting, LTD.

[3] For any one oil pollution incident, the Fund may pay up to $1 billion. Fund expenditures for natural resource damage assessments and claims in connection with a single incident are limited to $500 million of that $1 billion.

[4] BP established the trust under Delaware law, which generally provides that the principal of the trust can be used only for the purposes stated in the trust agreement and that the terms of the agreement cannot be modified and are legally enforceable by the trustees.

[5] GAO, *Deepwater Horizon Oil Spill: Preliminary Assessment of Federal Financial Risks and Cost Reimbursement and Notification Policies and Procedures*, GAO-11-90R (Washington, D.C.: Nov. 12, 2010).

[6] GAO, *Deepwater Horizon Oil Spill: Update on Federal Financial Risks and Claims Processing*, GAO-11-397R (Washington, D.C.: Apr. 18, 2011).

[7] Natural resource damage assessments are conducted to evaluate the nature and extent of injuries resulting from an oil spill, and to determine restoration actions needed to bring injured natural resources and services back to what they were prior to the incident.

[8] Contingent liabilities are potential liabilities that stem from an existing condition, situation, or set of circumstances involving uncertainty as to possible loss to an entity. The uncertainty will ultimately be resolved when one or more future events occur or fail to occur.

[9] Under OPA, a responsible party can also assert a defense to liability if the oil spill was caused solely by, among other things, an "act of God," an "act of war," the acts or omissions of an independent third party (provided certain conditions are satisfied), or any combination of these. 33 U.S.C. 2703. NPFC guidance acknowledges that terrorism or other criminal acts may present a defense to liability under OPA. NPFC, *NPFC User Reference Guide (eURG),* Appendix B, *Federal On-Scene Coordinator Funding Information for Oil Spills and Hazardous Materials Releases* (Washington, D.C.: April 2003). available at http://uscg.mil/npfc/URG/default.asp.

[10] Removal costs are incurred by the federal government or any other entity taking approved action to contain and clean up the spill.

[11] Users of offshore facilities on outer continental shelf lands have similar requirements to those covering vessels. Under OPA, they must submit evidence of an Oil Spill Financial Responsibility for an offshore facility (that generally is capable of discharging more than

1,000 barrels of oil) to the Department of Interior (DOI) and receive approval. DOI's regulations for the process are set out in 30 C.F.R. part 253.

[12] See appendix III for NPFC's individual and business claims process.

[13] 33 U.S.C. 2706.

[14] 15 C.F.R. 990.41; 15 C.F.R. 990.65.

[15] Affected areas may include both water and land resources, such as waterways or beaches.

[16] Appendix IV discusses NPFC's cost reimbursement process.

[17] If a claim were paid by the Fund, the Fund could recover administrative costs attributable to the claim under 33 U.S.C. 2715.

[18] 40 C.F.R. Part 300.

[19] Pub. L. No. 110-343, § 405, 122 Stat. 3765, 3860 (Oct. 3, 2008).

[20] The Federal On-Scene Coordinator has responsibility for overseeing oil spill response efforts and determining that efforts were conducted in accordance with the National Contingency Plan. To pay government agencies' oil spill removal costs, the Federal On-Scene Coordinator issues authorizations to quickly obtain services and assistance from government agencies, verifies that the services or goods were received and consistent with the National Contingency Plan, certifies the supporting documentation, and sends the cost documentation to NPFC, which authorizes the Coast Guard's Finance Center to pay the government agencies.

[21] OPA provides for the designation of federal, state, and if designated by the Governor of the state, local officials, to act on behalf of the public as trustees for natural resources. In addition, OPA provides for designations of Indian tribe and foreign officials to act as trustees for natural resources on behalf of, respectively, the tribe or its members and the foreign government.

[22] Pub. L. No. 107-295, § 323, 116 Stat. 2064, 2104 (Nov. 25, 2002).

[23] Pub. L. No. 111-191 § 1, 124 Stat. 1278 (June 15, 2010).

[24] Natural resource damage claims are claims for natural resource damages arising out of oil spills and include costs to restore, rehabilitate, replace, or acquire the equivalent of the injured resource, any interim lost use or diminution in value of the injured resource pending restoration, and the reasonable cost of assessing those damages.

[25] On May 11, 2010, NPFC notified BP and Transocean Holdings Incorporated that BP's advertisement of its claims process was sufficient, and Transocean should not advertise and should coordinate claims processing with BP. According to NPFC officials, they wanted to avoid public confusion and have only one responsible party advertise for claims.

[26] *Legal Liability Issues Surrounding the Gulf Coast Oil Disaster:* Hearing Before the H. Comm. on the Judiciary, 111th Cong. 85-92 (2010) (statement of Darryl Willis, Vice President, Resources, BP America).

[27] Under terms of the Trust, BP may be called upon to pay non-economic damages as a result of litigation or other settlements.

[28] For example, 46 U.S.C. § 30104, (commonly known as the Jones Act),establishes liability for injury or death of seamen incurred in the course of their employment. For additional information regarding the oil spill legal framework, see enclosure III in our November 2010 product (GAO-11-90R).

[29] On September 1, 2011, the GCCF closed 7 of the claims centers due to reduced activity and need for the centers.

[30] GCCF,*Gulf Coast Claims Facility Protocol for Interim and Final Claims* (Nov. 22, 2010).

[31] The current cap balance is calculated by NPFC and consists of the actual expenditures from the Fund and amounts obligated by NPFC but have yet to be expended from the Fund.

[32] BP has established collateral for the remaining portion of the $20 billion yet to be funded.

[33] NPFC has billed the responsible parties based on a combination of Coast Guard actual costs and a percentage of the amount NPFC has obligated for agencies through PRFAs and MIPRs.

[34] The United States has not encountered a spill comparable to the *Deepwater Horizon* oil spill since the 1989 *Exxon Valdez* spill, which reported that its natural resource damages were at least $1 billion (within the $1 billion cap the Fund has a $500 million statutory limit on natural resource damage assessments and claims). The responsible parties can pay natural resource damage assessment costs directly to the relevant federal agencies.

[35] Enclosure III of our November 2010 report (GAO-11-90R) discusses the legal framework of oil spills.

[36] The National Commission on BP Deepwater Horizon Oil Spill and Offshore Drilling was established through Exec. Order No. 13,543 (May 21, 2010) to examine the relevant facts and circumstances concerning the root causes of the *Deepwater Horizon* explosion and developed options to guard against, and mitigate the impact of, any oil spills associated with offshore drilling in the future. The Commission's responsibilities include recommending improvements to federal laws, regulations, and industry practices.

[37] 33 U.S.C. 1321(b) (6) and (7).

[38] GAO-11-90R.

[39] As of May 31, 2011, all 570 individual and business claims finalized by NPFC have been denied for reasons such as failure to demonstrate that damages were the result of the spill and lack of documentation.

[40] Pub. L. No. 110-343, div. B, tit. IV § 405(b), 122 Stat. 3765, 3861 (Oct. 3, 2008).

[41] In fiscal year 2010, the Fund received $475.9 million from the per barrel tax, $518.4 million from payments related to *Deepwater Horizon,* and $18.7 million from interest on investments.

[42] GAO, *Maritime Transportation: Major Oil Spills Occur Infrequently, but Risks to the Federal Oil Spill Fund Remain*, GAO-07-1085 (Washington D.C.: Sept. 7, 2007).

[43] The tax expired in December 1994 and was reinstated by the Energy Policy Act of 2005 beginning April 2006.

[44] GAO-07-1085.

[45] As of May 31, 2011, NPFC had finalized 570 claims. Our sample for claims testing was selected from the 432 claims finalized as of April 30, 2011.

[46] According to 33 C.F.R. 136.115(c), if NPFC fails to make final disposition of a claim within 6 months after it is filed, at the option of the claimant any time thereafter, be deemed a final denial of the claim.

[47] A request for reconsideration must be received by NPFC within 60 calendar days after the date the denial was mailed to the claimant or within 30 days after receipt of the denial by the claimant, whichever is earlier. 33 C.F.R. 136.115(d).

[48] The Coast Guard's Finance Center (FINCEN) is located in Chesapeake, Va. and serves as the data center for finance and procurement, central bill paying and financial accounting center for the Coast Guard.

[49] U.S. Coast Guard, NPFC, *Case Management Division Standard Operating Procedures, CM SOP Appendix,* NPFCINST M16451.23, April 2011.

[50] U.S. Coast Guard, NPFC, *Standard Operating Procedures of the Claims Adjudication Division,* NPFCINST M16451.21 (April 2004).

[51] GAO, *Standards for Internal Control in the Federal Government,* GAO/AIMD-00-21.3.1 (Washington, D.C.: November 1999).

[52] In response to the *Deepwater Horizon* oil spill, on May 2010 DHS's Federal Emergency Management Agency (FEMA) was tasked as the coordination lead for the Deepwater Integrated Services Team. The focus of the IST was to monitor BP's claims process and coordinate the delivery of federal programs that could provide social services and small business assistance for individuals, families, and businesses, as well as state and local government entities affected by the spill.

[53] OMB Memorandum M-10-29, *Identifying and Documenting Costs of Government Activities Related to the BP Deepwater Horizon Oil Spill* (July 1, 2010); OMB Management Procedures Memorandum No. 2010-35, *Reporting Costs of Government Activities Related to the BP Deepwater Horizon Oil Spill* (Oct. 5, 2010); OMB Memorandum M-11-09, *Supplemental Guidance on Reporting Costs of Government Activities Related to the BP Deepwater Horizon Oil Spill* (Jan. 12, 2011); and, OMB Memorandum M-11-09 (revised), *Supplemental Guidance on Reporting Costs of Government Activities Related to the BP Deepwater Horizon Oil Spill* (Jan. 13, 2011).

[54] Examples included NOAA serving as the lead science agency and Department of Energy evaluating methods and risks to stem the flow of oil.

[55] According to Justice officials, the $81.6 million payment will be deposited into the U.S. Treasury.

[56] The National Center for Disaster Fraud was established by the Criminal Division of the United States Department of Justice in the fall of 2005 in the wake of Hurricane Katrina. The Center is located in Baton Rouge, Louisiana. Its purpose is to receive and screen reports from the public about possible fraud relating to disasters of all types, and to refer those reports to the field offices of appropriate federal law enforcement agencies.

[57] Numerous individuals, businesses, states, and the federal government have begun various actions in a number of courts against several companies, including BP, seeking damages or declaratory or injunctive relief under several laws, including OPA. Many of these pending cases have been consolidated in multidistrict litigation in the U.S. District Court for the Eastern District of Louisiana. See http://www.laed.uscourts.gov/OilSpill/OilSpill.htm.

[58] If a claimant decides to commence litigation against the responsible parties, NPFC will not review the same claim until the litigation has concluded.

[59] GAO-11-90R, 34.

End Notes for Appendix I

[1] GAO-11-90R.

End Notes for Appendix II

[1] See GAO, *Maritime Transportation: Major Oil Spills Occur Infrequently, but Risks to the Federal Oil Spill Fund Remain*, GAO-07-1085 (Washington, D.C.: Sept. 7, 2007).

[2] GAO, *Standards for Internal Control in the Federal Government*, GAO/AIMD-00-21.3.1 (Washington, D.C.: November 1999).

[3] FINCEN is located in Chesapeake, Va., and serves as the data center for finance and procurement, central bill paying, and financial accounting for the U.S. Coast Guard.

[4] The 954 *Deepwater Horizon* disbursements were identified by retaining only transactions with the *Deepwater Horizon* Federal Project Number (N10036), were expenditure-type transactions ("EXP") and had a document ID identifying them as either a Military Interdepartmental Purchase Request (MIPR) (doc ID "28") or Pollution Removal Funding Authorization (PRFA) (doc ID "34") disbursement, and eliminating the Treasury confirmations ("JE Category field"). Federal Project Numbers are unique numbers assigned by NPFC to identify oil pollution incidents.

[5] NPFC's Claims Processing System is a work flow system that supports the initial receipt, administrative processing, and subsequent routing and payment of claims for NPFC.

[6] When information of an oil spill is received, the source or sources of the discharge or threat are designated where possible and appropriate. If the designated source is a vessel or facility, the responsible party and the guarantor, if known, are notified by telephone, telefax, or other rapid means of that designation. The designation will be confirmed by a written Notice of Designation.

[7] As of May 31, 2011, Justice had sent one bill in the amount of $81.6 million for cost recovery from the responsible parties.

End Notes for Appendix III

[1] To facilitate providing states with funds quickly for their oil spill response costs, NPFC has developed an expedited claims procedure for state governments.

[2] Date of completion of all removal actions is defined as the actual date of completion of all removal actions for the incident or the date the Federal On-Scene Coordinator determines that the removal actions which form the basis for the costs being claimed are completed, whichever is earlier.

[3] Letters of Federal Interest are issued by the Federal On-Scene Coordinator to assert the need for positive responsible party action.

[4] NPFC's Case Management Division is responsible for providing access to the Emergency Fund for Federal removal costs and for accurate cost documentation to support cost recovery. NPFC's Claims Adjudication Division is responsible for providing assistance to the victims of oil spills by receiving, processing, adjudicating, settling, and approving the payment of OPA claims. It is also responsible for advertising for claims if the responsible party does not.

[5] To begin the claims process, 33 U.S.C. 2714 provides that once an incident becomes known, the source or sources of a discharge or threat shall be designated where possible and appropriate. And, if the designated source is a vessel or a facility, the responsible party and the guarantor, if known, shall be immediately notified of the designation. "Designation" is an OPA term used in connection with the initiation of the claims process and is aimed at the advertisement of responsible party responsibility to potential claimants.

[6] U.S. Coast Guard, NPFC, *Standard Operating Procedures of the Claims Adjudication Division*, NPFCINST M16451.21 (April 2004).

[7] NPFC's Natural Resource Damage Claims Division adjudicates claims for natural resource damages arising out of oil spills (or the substantial threat of a spill) to the navigable waters of the United States. Those damages may include the cost to restore, rehabilitate, replace, or

acquire the equivalent of the injured resource; any interim lost use or diminution in value of the injured resource pending restoration; and, the reasonable cost of assessing those damages.

[8] A major difference between NPFC's Natural Resource Damage Claims Division and its Claims Adjudication Division is that the Natural Resource Damage Claims Division establishes Interagency Agreements between NPFC and the Federal Natural Resource Damage Trustees to fund initiation of natural resource damage assessments, whereas the Claims Adjudication Division does not.

End Notes for Appendix IV

[1] Private companies, contractors, and other nongovernmental entities may also be part of a coordinated spill response. Contractors performing response or removal actions on behalf of the Federal On-Scene Coordinator are eligible to receive reimbursement from the Fund. However, since the *Deepwater Horizon* response efforts predominantly have involved federal and state governmental entities, this cost reimbursement appendix will focus on the process for reimbursing government agencies.

[2] The National Response Center, located at U.S. Coast Guard Headquarters, is the national communications center, continuously manned for handling activities related to response actions. The Center acts as the single point of contact for all pollution incident reporting.

[3] The National Contingency Plan requires that oil releases are reported to the National Response Center, which is staffed by the Coast Guard.

[4] DOD uses MIPRs in order to obtain and provide services to agencies.

[5] FINCEN is located in Chesapeake, Va., and serves as the data center for finance and procurement, central bill paying, and financial accounting for the U.S. Coast Guard.

End Notes for Appendix VI

[1] Department of Commerce Inspector General, Final Memorandum No. OIG-11-016-M *Survey of National Oceanic and Atmospheric Administration's (NOAA) System and Processes for Tracking Oil Spill Costs* (December 2010).

End Notes for Appendix VII

[1] The seven federal agencies that have been reimbursed from the Fund are the Departments of Defense, Interior, Commerce, Homeland Security, Energy, Agriculture, and the Environmental Protection Agency.

[2] The six federal agencies that have intergovernmental agreements in place, but have not been paid from the Fund, are the Departments of Health and Human Services, Labor, Justice, the National Transportation Safety Board, National Security Agency, and the Office of the Director of National Intelligence.

In Deepwater Horizon: ISBN: 978-1- 61942-499-9
Editors: D. A. Michaels and G. S. Aston © 2012 Nova Science Publishers, Inc

Chapter 2

DEEPWATER HORIZON OIL SPILL: UPDATE ON FEDERAL FINANCIAL RISKS AND CLAIMS PROCESSING*

United States Government Accountability Office

Washington, DC 20548
April 18, 2011
Congressional Requesters

On April 20, 2010, an explosion occurred on BP America Production Company's (BP) leased mobile offshore drilling unit *Deepwater Horizon*. The total cost to clean up the massive and unprecedented oil spill in the Gulf of Mexico following the *Deepwater Horizon* explosion (including costs to help pay for the spill's adverse impact on businesses and individuals in the region) are yet unknown, but have been estimated in the tens of billions of dollars. The extent to which the federal government will ultimately be required to pay costs associated with the *Deepwater Horizon* oil spill remains unclear.

The complex legal framework in place for oil spill liability and response funding will play an integral role in determining who is responsible and will ultimately pay the costs associated with the *Deepwater Horizon* oil spill. In this regard, the Oil Pollution Act of 1990,[1] as amended (OPA), which

* This is an edited, reformatted and augmented version of the United States Government Accountability Office publication, GAO-11-397R, dated April 18, 2011.

Congress enacted after the *Exxon Valdez* spill in 1989, authorized use of the Oil Spill Liability Trust Fund (Fund) to pay for certain oil spill cleanup costs and damages using federal tax revenues for immediate response costs and when the responsible parties cannot be identified or do not pay. OPA also provided that the federal government may subsequently seek reimbursement for these costs from responsible parties.[2] The Fund, which is administered by Coast Guard's National Pollution Funds Center (NPFC), is subject to a $1-billion cap on the total amount of expenditures per incident.

NPFC designated two BP subsidiaries—BP Exploration and Production and its guarantor, BP Corporation North America, Inc.—and five other companies as responsible parties for *Deepwater Horizon* oil spill related claims. Shortly after the spill, at the direction of NPFC, BP began to receive and process all claims against responsible parties. In June 2010, at the urging of the White House and Department of Justice, BP established a new claims processing facility—the Gulf Coast Claims Facility (GCCF). GCCF began operations on August 23, 2010, and is responsible for handling claims from individuals and businesses for damages resulting from the *Deepwater Horizon* oil spill. For those claims submitted to the GCCF that are rejected or not paid within 90 days, claimants may file OPA-compensable claims with NPFC to request reimbursement from the Fund.

BP also established an irrevocable trust (Trust), to which BP is to provide a total of $20 billion by 2014, primarily for the purpose of paying GCCF and other claims related to the *Deepwater Horizon* oil spill.[3] The Trust is to pay some OPA-compensable claims and some other claims for personal injuries that are not OPA-compensable, but for which BP would be liable under other federal or state laws, such as the Jones Act or state oil pollution acts.[4]

In November 2010,[5] we reported on our preliminary assessment of the potential financial risks to the federal government associated with the *Deepwater Horizon* oil spill cleanup costs. The attached briefing provides information updated since our preliminary assessment. For this briefing our objectives are to provide updated information on (1) the financial risks to the federal government associated with the cap on expenditures from the Fund and (2) claims submitted to and reviewed NPFC and GCCF, and those paid by GCCF. We also provide an update of the status of agency actions to respond to the recommendations made in our November 2010 report. This is the second in a planned series of three reports on our work in this area.

Our third report, planned for the summer of 2011, is intended to be a capping report with an updated assessment of: (1) the financial risks to the federal government associated with the Fund; (2) NPFC Fund cost reimbursements and claims and related processes; and (3) the federal framework for monitoring and oversight of responsible parties' actions to pay costs associated with the *Deepwater Horizon* oil spill.

Scope and Methodology

To provide an update on the financial risks to the federal government and the Fund, we obtained and summarized available data from NPFC on obligated and actual costs incurred and reviewed publicly available financial information of responsible parties through March 2011. We also obtained updated data on reported costs incurred in relation to the cap on expenditures from the Fund.

In order to update information about claims submitted and reviewed by NPFC and GCCF, we used available NPFC and GCCF claims data through March 2011, to describe the number and types of claims filed by individuals and businesses against the Trust and the Fund, and the number and dollar amounts claimed, reviewed, and paid. We also obtained information on NPFC's claims contingency planning for handling potential surges in claims submitted related to the *Deepwater Horizon* oil spill. We also obtained information from NPFC officials about the status of the recommendations made in our November 2010 report.

We conducted our work from November 2010 to March 2011 in accordance with all sections of GAO's Quality Assurance Framework that are relevant to our objectives. The framework requires that we plan and perform the engagement to meet our stated objectives and that we discuss any limitations in our work. We believe that the information and data obtained, and the analysis conducted, provide a reasonable basis for our findings and conclusions.

We performed limited procedures to determine the Fund expenditures reported were reasonable for our reporting purposes.

Results in Brief

With reported Fund costs of about $629.5 million as of March 31, 2011, NPFC had obligated or incurred costs that could result in over 60 percent of the amount available under the Fund's statutory $1-billion-per-incident-expenditure-cap.[6] If, regardless of any reimbursements from responsible parties, total Fund expenditures exceed the $1-billion cap, agencies may be required to rely on reallocating their appropriated funding to cover costs they incur or obtain supplemental funding. In addition, agencies may be unable to cover some of their costs and NPFC would be unable to pay any additional claims to individuals and businesses related to the *Deepwater Horizon* oil spill. We are reiterating our prior matter that Congress should consider changing the calculation of expenditures made against the Fund's $1-billion- per-incident-expenditure-cap to take into account reimbursements from responsible parties. Ultimately, the federal government's financial risk will continue to be closely linked with actions taken by the responsible parties to pay such costs. To date, BP has continued to fund the Trust established in August 2010 to pay for *Deepwater Horizon* oil spill claims as agreed.

With respect to claims processing, NPFC has taken a number of steps to monitor the GCCF's claims processing in planning for contingencies to help ensure it can effectively process any future surges in the number of claims it receives as a result of rejected GCCF claims that NPFC may receive for adjudication related to the *Deepwater Horizon* oil spill. These actions helped NPFC to process a sharp increase in the number of claims that individuals and businesses submitted to NPFC in December 2010. NPFC officials told us they monitor ongoing GCCF activities in order to forecast and take actions to mitigate potential surges in the number of claims that may come to NPFC for adjudication. As of March 2011, GCCF had established four types of claims payments—Emergency Advanced Payments, Quick Payments (Final), Interim Payments, and Full Review Payments (Final). As of March 31, 2011, GCCF had paid approximately $3.7 billion on 281,308 claims and denied over 4,000 claims.

In response to our previous recommendations, NPFC reported that it plans to update its policies and procedures in August and October 2011 to address three of the four recommendations made in our November 2010 report. These recommendations were directed at helping NPFC establish and maintain effective cost reimbursement policies and procedures for the Fund and update NPFC's current policies to reflect current organization, structure and management's directives. For the other recommendation, NPFC contends that its current procedures, which allow for invoices sent to responsible parties to serve as notification for cost recovery, provide adequate documentation of responsible party designation. However, NPFC intends to review, clarify, and update its designation procedures by October 31, 2011. We believe that clarification of this process is necessary. As stated in our November 2010 report, we found that NPFC's existing procedures for notifying responsible parties, including the use of an invoice as notification of "responsible party" designation, were not clear. For example, NPFC sent an invoice for reimbursement to one of the *Deepwater Horizon* oil spill responsible parties that it considered as formal notification of the entity's financial responsibilities. However, a representative of the entity later publicly stated it had not received notification of a "responsible party" designation.

Agency Comments and Our Evaluation

We provided a draft of our briefing to the Department of Homeland Security's (DHS) and the Department of Justice's (DOJ) management for comment. DHS commented that NPFC continues to disagree with what it understands to be GAO's concern that a notice of designation of a discharge source issued to some responsible parties, but not to all responsible parties that are eventually identified, risks confusion and breakdowns in the claims management and cost reimbursement process. However, it also commented that NPFC intends to review, clarify, and update its designation procedures on or by October 31, 2011. DOJ commented that GAO has not identified any statutory responsibility that NPFC failed to fulfill under OPA Title 1, nor identified any policy basis for our responsible party notification recommendation.

We did not assert that NPFC did not comply with law or policy. Nonetheless, as we previously reported, clarification of responsible party designation procedures is necessary to avoid possible confusion over responsibility and breakdowns in the claims management and cost reimbursement process. Consequently, we are encouraged that NPFC intends to update its procedures to clarify responsible party designations.

Susan Ragland
Director
Financial Management and Assurance

List of Requesters

The Honorable John Conyers, Jr.
Ranking Member

Committee on the Judiciary
House of Representatives

The Honorable Bennie G. Thompson
Ranking Member
Committee on Homeland Security
House of Representatives

The Honorable Tom Carper
Chairman
Subcommittee on Federal Financial Management,
Government Information, Federal Services, and International Security
Committee on Homeland Security and Governmental Affairs
United States Senate

The Honorable Sheldon Whitehouse
Chairman
Subcommittee on Oversight
Committee on Environment and Public Works United States Senate

The Honorable Mary Landrieu
Chairwoman
Subcommittee on Disaster Recovery Committee on
Homeland Security United States Senate

The Honorable Michael C. Burgess

The Honorable Nick J. Rahall, II
House of Representatives

ENCLOSURE

Introduction

On April 20, 2010, an explosion occurred on BP America Production Company's (BP) leased mobile offshore drilling unit, *Deepwater Horizon*. The total cost to clean up the massive and unprecedented oil spill in the Gulf of Mexico following the *Deepwater Horizon* explosion, as well as costs to help pay for the spill's adverse impact on businesses and individuals in the region, are not yet known, but have been estimated in the tens of billions of dollars. Further, the extent to which the federal government will ultimately be required to pay any of the costs associated with the *Deepwater Horizon* oil spill remains unclear.

In November 2010,[1] we reported on our preliminary assessment of the potential financial risks to the federal government associated with the *Deepwater Horizon* oil spill cleanup costs. The overall objectives for our work in this area are to assess (1) financial risks to the federal Oil Spill Liability Trust Fund (Fund) and the federal government as a result of the *Deepwater Horizon* oil spill, (2) the Coast Guard's National Pollution Funds Center's (NPFC) cost reimbursement and claims policies and procedures for *Deepwater Horizon* oil spill costs, and (3) the framework for federal monitoring and oversight efforts over the responsible parties for the *Deepwater Horizon* oil spill, including federal efforts to oversee BP's and the Gulf Coast Claims Facility's (GCCF) *Deepwater Horizon* oil-spill claims payments.

Objectives

This briefing provides information updated since our preliminary assessment in November 2010. For this briefing, our objectives are to provide updated information related to the *Deepwater Horizon* oil spill on:

1) the financial risks to the federal government associated with the cap on expenditures from the Fund and
2) claims submitted to and reviewed by NPFC and GCCF, and those paid by GCCF.

We also provide an update of the status of agency actions to respond to the recommendations made in our November 2010 report.

This is the second in a planned series of three reports on our work in this area.

Our third report, planned for the summer of 2011, is intended to be a capping report addressing our overall objectives.

Scope and Methodology

To provide an update on the financial risks to the federal government, we obtained and summarized available data from NPFC on obligated and actual costs incurred and reviewed publicly available financial information of responsible parties through March 2011. We also obtained updated data on reported costs incurred in relation to the cap on expenditures from the Fund.

In order to update information about claims submitted and reviewed by NPFC and GCCF, we used available NPFC and GCCF claims data through March 2011, to describe the number and types of claims filed by individuals and businesses against the Trust and the Fund, and the number and dollar amounts claimed, reviewed, and paid. We also obtained information on NPFC's claims contingency planning for handling potential surges in claims submitted related to the *Deepwater Horizon* oil spill.

We also obtained information from NPFC officials about the status of the recommendations made in our November 2010 report.

We conducted our work from November 2010 to March 2011 in accordance with all sections of GAO's Quality Assurance Framework that are relevant to our objectives. The framework requires that we plan and perform the engagement to meet our stated objectives and that we discuss any limitations in our work. We believe that the information and data obtained, and the analysis conducted, provide a reasonable basis for our findings and conclusions.

We performed limited procedures to determine the Fund expenditures reported were reasonable for our reporting purposes.

Background

A complex landscape of laws and regulations governs the liability for oil spill costs of different parties. Injuries and damages that arise from an oil spill incident are governed by federal statutes and common law, federal securities laws, and various state laws. For example, the Oil Pollution Act of 1990 (OPA),[2] as amended, places the primary liability for the cost of the oil spills—up to certain limits—on the responsible party or parties for removal costs and damages specified in OPA (referred to as OPA-compensable damages). OPA authorizes the use of the Fund, which NPFC administers, for federal cleanup and natural resource restoration. NPFC monitors the sources and uses of the Fund, adjudicates claims submitted to the Fund for payment, and pursues reimbursements from responsible parties for costs and damages paid by the Fund and certain other recoverable costs.

Under OPA, the authorized limit on expenditures to be paid from the Fund is currently $1-billion in total expenditures per incident, with a concurrent limit of $500 million per incident for natural resource damage assessments and claims.

Following the *Deepwater Horizon* oil spill, the Coast Guard, without in any way relieving other responsible parties of liability, directed BP to establish a single claims facility for all responsible parties to centralize claims processing for claimants.[3] In June 2010, at the urging of the White House and the Department of Justice, BP established a new claims processing facility—GCCF— and announced creation of a $20 billion escrow account (Trust) to satisfy claims resolved by GCCF and certain other claims, including natural resource damages. BP has also pledged collateral to secure its obligation to contribute the full $20 billion to the Trust.

BP established GCCF to provide a mechanism for individuals and businesses to file claims for costs and damages incurred as a result of the *Deepwater Horizon* oil spill. Because NPFC bills the responsible parties directly for costs agencies have incurred in response to the *Deepwater Horizon* oil spill, BP pays these costs and they are not paid from the Trust. Payments received by NPFC from BP are deposited into the Fund.

Individuals and businesses are required to first file with GCCF for *Deepwater Horizon*-related oil spill claims. For those claims submitted to GCCF that are rejected or not paid within 90 days, claimants may file OPA-compensable claims with NPFC to request reimbursement from the Fund.

GCCF began operations and started accepting claim forms on August 23, 2010. GCCF, administered by Kenneth R. Feinberg, draws funds from the Trust to pay claims. The payments are intended to provide compensation for both OPA-compensable and certain non-OPA-compensable claims.

BP established an irrevocable Trust (for the announced escrow account) on August 6, 2010, designating three trustees[4] with fiduciary responsibility to collect promised contributions from BP and make disbursements to permitted categories of beneficiaries. BP committed to fund the Trust on a quarterly basis over 3-1/2 years for a total of $20 billion to be paid into the Trust as of 2014.[5] The Trust is to pay some OPA-compensable claims and some other claims for personal injuries that are not OPA-compensable, but for which BP would be liable under other federal or state laws, such as the Jones Act or state oil pollution acts.[6]

Results in Brief

With reported Fund costs of about $629.5 million as of March 31, 2011, NPFC has incurred costs that could result in payments of over 60 percent of the funds available under the Fund's statutory $1-billion-per-incident-expenditure-cap.[7] If total Fund expenditures for the *Deepwater Horizon* oil spill exceed $1 billion, agencies may be required to rely on reallocating their appropriated funding to cover costs they incur or obtain supplemental funding. In addition, agencies may be unable to cover some costs and NPFC would be unable to pay any additional claims related to the *Deepwater Horizon* oil spill. Ultimately, the federal government's financial risk will continue to be closely linked with actions taken by BP and the other responsible parties to pay claims and other costs. To date, BP has continued to meet its stated commitment to pay the costs associated with the *Deepwater Horizon* oil spill and has continued to fund the August 2010 agreement establishing a $20 billion Trust that GCCF can draw from to pay claims.

With respect to claims processing, NPFC actions are necessarily closely tied to those of GCCF. NPFC has taken a number of steps to monitor GCCF's claims processing in planning for contingencies to help ensure it can effectively process any future surges in the number of claims it receives as a result of rejected GCCF claims that NPFC may receive for adjudication related to the *Deepwater Horizon* oil spill. These actions helped NPFC to process a sharp increase in the number of claims individuals and businesses submitted to NPFC in December 2010.

On our previous recommendations, NPFC reported that it plans to update its policies and procedures in August and October 2011 to address three of the four recommendations made in our November 2010 report. For the other recommendation, NPFC contends that its current procedures, which allow for invoices sent to responsible parties to serve as notification for cost recovery, provide adequate documentation and notification of responsible party designations. We disagree. As stated in our November 2010 report, we found that NPFC's existing procedures for notifying responsible parties using invoices did not clearly communicate their "responsible party" designation. For example, an official from a *Deepwater Horizon* oil spill responsible party who had received an invoice from NPFC, stated during a July 2010 hearing that his company had not received notification of designation. Also, NPFC sent the notice to Transocean Holdings Incorporated, but Transocean replied that the correct entity is Transocean Holdings, LLC.

Risk That Total Expenditures from the Fund Will Reach the Cap

Federal agencies continue to incur *Deepwater Horizon*-related removal and other costs. As NPFC continues to make expenditures from the Fund to reimburse federal agency costs and directly pay for other *Deepwater Horizon*-related costs, NPFC reported, as of March 31, 2011, it has incurred costs that could result in payments of over 60 percent of the funds available under the $1-billion-per-incident-expenditure-cap from the Fund. Once the expenditures reach the cap, NPFC will be precluded from making any additional cost reimbursements to agencies or paying any additional claims related to the *Deepwater Horizon* oil spill.

According to internal Coast Guard reports, as of March 31, 2011, NPFC had obligated or incurred approximately $629.5 million against the Fund toward the $1-billion-per-incident-expenditure-cap.[8] This amount includes:

- $451.4 million obligated against the Fund to reimburse government agencies' Pollution Removal Funding Authorizations (Federal Authorizations) and Military Interdepartmental Purchase Requests (MIPR). Of the total obligated amount, as of March 31, 2011, NPFC had approved about $193.9 million for payment.[9]
- $130.2 million expended by the Coast Guard and charged directly to the Fund. These costs are referred to by the Coast Guard as direct costs and include contracts and travel directly related to the oil spill response.[10]
- $47.8 million obligated against the Fund for the initiation of natural resource damage assessments (NRDA) to the Department of the Interior (DOI).[11]

NPFC officials told us that they have not estimated a time frame for when they anticipate the cap will be reached. However, they stated that there is a significant risk the cap could be reached in fiscal year 2011 as agencies continue to conduct significant removal activities related to the *Deepwater Horizon* oil spill.

If expenditures from the Fund collectively exceed the $1-billion-per-incident-cap, agencies and claimants could no longer receive reimbursement from the Fund. In that event, federal agencies might have to turn to options such as requesting supplemental appropriations or reallocating funds from their annual appropriations or using other agency budgetary resources to cover costs that would otherwise be reimbursed by the Fund. Further, if agencies stopped funding *Deepwater Horizon* oil-spill-related activities, this could affect the federal government's ability to complete oil spill removal and other related efforts.

Ultimately, the federal government's financial exposure will continue to be closely linked with BP and the other responsible parties' actions concerning the *Deepwater Horizon* oil spill. BP has committed to paying costs for the *Deepwater Horizon* oil spill including reimbursing the Fund for its *Deepwater Horizon*-related expenditures, even to the extent such costs exceed the $20 billion it has agreed to set aside. However, circumstances may occur that adversely impact BP or other responsible parties' financial condition or ability (above BP's collateralized pledge) to pay such claims including reimbursing the Fund for *Deepwater Horizon* costs paid.

Through March 2011, BP has continued to fund the Trust established in August 2010 to pay for *Deepwater Horizon* oil spill claims as agreed. BP committed to fund a $20 billion irrevocable Trust on a quarterly basis over 3-1/2 years. As of March 31, 2011, BP has made the required payments that total $6.25 billion. In addition, as shown on table 1, GCCF has paid approximately $3.7 billion on over 280,000 claims from the Trust as of March 31, 2011.

Claims-Processing Status Update

Table 1. Claims Paid by GCCF as of March 31, 2011 (unaudited)

Dollars in millions		
Type	Number of claims paid	Amount
Emergency Advanced Payments	169,005	$2,580.2
Interim Payments	4,313	48.4
Quick Pay (Final) a	101,474	947.0
Full Review (Final)a	6,516	79.0
Totalb	281,308	$3,654.6

Source: GAO analysis of GCCF data.

[a] As of March 31, 2011, 38 percent of the payments from GCCF were paid as either Quick Pay or Full Review, both of which require these claimants to sign a release waiving any rights they may have against responsible parties to file or participate in legal action, or to submit any claim to NPFC for payment.

[b] As described in our November 2010 report, claims approved by GCCF are paid from a Trust established and funded (up to $20 billion) by BP. Prior to the establishment of GCCF, BP had received and directly paid claims from individuals and businesses totaling $396.0 million.

As of March 2011, while GCCF has established four types of claim payments and paid over $3.6 billion.

- Emergency Advanced Payments. Payments that were available to individuals and businesses that experienced financial hardship resulting from damages incurred from the *Deepwater Horizon* oil spill and filed claims by November 23, 2010.[12]
- Quick Pay (Final). Payments to a claimant who has been paid an Emergency Advance Payments by GCCF which require the claimant to sign a release[13] and within 14 days be paid $5,000 if an individual claimant or $25,000 if a business claimant without having to submit additional supporting documents or go through further claims review.
- Interim Payments. Payments for documented past damages caused by the *Deepwater Horizon* oil spill. The Interim Payments will not compensate for future losses or damages.
- Full Review (Final). Payments for all past and future losses caused by the *Deepwater Horizon* oil spill. Claimants who accept a final payment are required to sign a release.[13]

With respect to claims processing, NPFC has taken a number of steps in planning for contingencies to help ensure it can effectively handle any surges in the number of claims it receives for adjudication as a result of rejection from GCCF related to the *Deepwater Horizon* oil spill. NPFC officials told us they monitor ongoing GCCF activities in order to forecast and take actions to mitigate potential surges in the number of claims that may come to NPFC for adjudication.

NPFC has established a contract through October 31, 2011, for additional claims reviewers.[14] The contract states that the contractor-provided services should allow NPFC management to make decisions based on the contractor's review. It also states that in all cases, NPFC is the final adjudicator on all claims.

In addition, NPFC officials told us that NPFC has plans to augment its claims division using Coast Guard reservists and could also reassign NPFC staff as needed to assist in the claims adjudication process.

To date, NPFC's actions have enabled it to manage its claims processing workloads. For example, in December 2010, NPFC experienced a surge in claims after 90 days had elapsed from receipt of many initial claims by GCCF,[15] and the deadline for submitting certain claims to GCCF had passed.

NPFC claims data showed that the number of monthly claims submitted to NPFC for the *Deepwater Horizon* oil spill significantly increased during the period from November 2010 to January 2011. (See Figure 1.) According to an NPFC official, this increase may be attributed to:

- Passage of GCCF's November 23, 2010, deadline for submitting Emergency Payment Claim applications;
- Public announcements made by GCCF's Administrator that GCCF was trying to clear its Emergency Payment Claims backlog by December 15, 2010; and
- November 23, 2010, was 90 days from when GCCF started accepting claims and in accordance with NPFC policies, if after 90 days the claim is without resolution, claims can be submitted to NPFC.

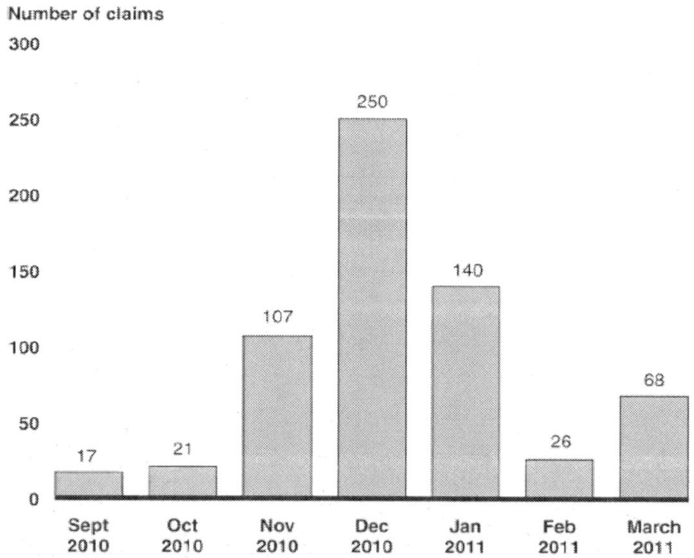

Note: NPFC first began receiving *Deepwater Horizon* oil spill-related claims in September 2010. The claims presented to NPFC through March 31, 2011, totaled $186 million.

Figure 1. Total Number of *Deepwater Horizon* Oil Spill Claims Presented to NPFC (unaudited).

The potential for another increase in the number of claims presented for payment to NPFC may occur again if a large number of claimants who are denied payments by GCCF choose to file their claims with NPFC at, or about, the same time. As of March 31, 2011, according to GCCF's Web site,[16] GCCF had, in addition to paying over 281,000 claims, denied over 4,000 claims from individuals and businesses, and issued determination letters that found over 3,000 claimants suffered no loss.

As of March 31, 2011, GCCF had more than 100,000 claims under review with additional claims being submitted daily. Among the claims under review, GCCF indicated that about 39,000 claims require additional information in order to be processed. Claimants who are denied payment by GCCF or whose claims are not settled within 90 days may pursue the following four options:

- Appeal GCCF's decision, if the claim is in excess of $250,000 under procedures established by GCCF Administrator;
- Commence litigation against the responsible parties in court;[17]
- File a claim with NPFC;[18] or
- Do nothing.

NPFC's March 31, 2011, data showed that since the *Deepwater Horizon* oil spill occurred in April 2010 it had received over 629 claims, totaling $186 million from individuals and businesses for this spill. NPFC had issued determinations for more than 538 of these claims (for about $163 million), all of which were denials.

NPFC denied the claims for the following reasons:

- Failure to prove damages were the result of the spill (39 percent)
- Lack of documentation (34 percent)
- Failure to prove damages (10 percent)
- Paid or being paid by responsible party (9 percent)
- Withdrawn by Claimant (5 percent)
- Not compensable under OPA, and therefore, not payable from the Fund (2 percent)
- Fraud (1 percent)

It is unclear at this time, if any, and if so how many, of the over 100,000 claims pending with GCCF, as of March 31, 2011, will ultimately result in claimants filing a claim with NPFC.

Since we last reported in November 2010, GCCF has updated its payment options and the deadlines associated with those options. GCCF's decisions and related actions affect the number of claims submitted to NPFC. When the deadlines were reached for the Emergency Advanced Payments claims to GCCF, a surge of claims were subsequently submitted to NPFC in December 2010.

Status of Prior Recommendations
Department of Homeland Security

NPFC has stated that actions are underway to address three of the four recommendations we made in our November 2010 report. Table 2 provides a status on the recommendations.

CONCLUSION

In our November 2010 report we offered a matter for congressional consideration that the Congress may want to consider setting a Fund cap per incident based upon net expenditures (expenditures less reimbursements). Since that report, expenditures have continued to be paid from the Fund and continue to approach the legislated cap. Given the risk that total expenditures from the Fund may reach the currently legislated cap, we are reiterating our prior suggestion that the Congress should consider amending the Oil Pollution Act of 1990, as amended (OPA),[19] or enacting new legislation that changes the calculation of expenditures made against the Fund's $1-billion-per-incident-expenditure-cap to take into account reimbursements from responsible parties.

We continue to believe that NPFC's procedures for identifying and documenting responsible party notification needs to be updated to clearly indicate the required mechanism to be used to notify responsible parties. Related to the *Deepwater Horizon* oil spill, NPFC sent formal letters of designation to some, but not all, of the responsible parties it identified. It provided only invoices to some responsible parties that reflect NPFC's assessment of liability for removal costs. NPFC's existing procedures for notifying responsible parties using invoices did not clearly communicate their "responsible party" designation. Consequently, we reiterate our previous recommendation that NPFC update its procedures to clearly indicate the required mechanism to be used to notify responsible parties for the *Deepwater Horizon* oil spill and future spills, in order to avoid confusion on whether an entity has financial responsibility for payment of oil-spill-related costs.

Table 2. Status of Prior Recommendations

Prior recommendation	Status
In order to help establish and maintain effective cost reimbursement policies and procedures for the Fund, we recommended that the Secretary of Homeland Security direct the Director of the Coast Guard's NPFC to update NPFC's policies and procedures to include:	
1. Current Fund reimbursement-billing practices that reflect both a percentage of federal agencies' obligations as well as expenditures, and	NPFC officials acknowledged that the billing practices for *Deepwater Horizon* are not documented in the agency's policies and procedures. NPFC officials told us they plan to formally incorporate the practices into its policies and procedures by October 31, 2011.
2. Specific procedural guidance on processing DOD requests for reimbursement using Military Interdepartmental Purchase Requests.	NPFC officials told us they plan to formally incorporate the procedures into NPFC's policies and procedures by October 31, 2011.
In order to ensure that responsible parties are properly notified of their responsibilities for an oil spill, we recommend that the Secretary of Homeland Security direct the Director of NPFC to	
3. Update NPFC's current policies to reflect current organization and structure and management's directives, and	NPFC officials stated that NPFC's current policies will be updated to reflect current organization and structure and management's directives by August 31, 2011.
4. Update NPFC's current procedures to provide detailed guidance and procedures for identifying and documenting responsible party notification.	NPFC officials disagreed with our recommendations and stated its responsible party designations are unrelated to the imposition of liability under OPA and that they serve the purpose of getting a responsible party to advertise the *Deepwater Horizon* oil spill claims process. NPFC's procedures provide that responsible parties and their guarantors are to be notified of their oil spill-related responsibilities. In accordance with its current procedures, NPFC sent formal letters of designation to some, but not all, of the responsible parties it identified for the *Deepwater Horizon* oil spill. To other responsible parties, NPFC provided only invoices that reflected NPFC's assessment of liability for removal costs. NPFC's existing procedures for notifying responsible parties using invoices did not clearly communicate their "responsible party" designation.
	We continue to believe that NPFC's procedures for identifying and documenting responsible party notification needs to be updated to clearly indicate the required mechanism used to identify and notify responsible parties of their financial obligations related to oil spills, including *Deepwater Horizon.*

Agency Comments and Our Evaluation

We provided a draft of our briefing to the management of the Departments of Justice and Homeland Security for comment. We received comments from the Department of Homeland Security on this briefing which stated NPFC continues to disagree with what it understands to be GAO's concern that a notice of designation of a discharge source issued to some responsible parties, but not to all responsible parties that are eventually identified, risks confusion and breakdowns in the claims management and cost reimbursement process. However, it also commented NPFC intends to review, clarify, and update its designation procedures on or by October 31, 2011. The Department of Justice also provided comments. In regards to our recommendation that NPFC update its procedures to provide detailed guidance and procedures for identifying and documenting responsible party notification, DOJ stated that GAO has not identified any statutory responsibility that NPFC failed to fulfill under OPA Title 1, nor has GAO identified any policy basis for this recommendation.

We did not assert that NPFC did not comply with legal requirements or policy concerning responsible party designations. Nonetheless, as we previously reported, clarification of responsible party designation procedures is necessary to avoid possible confusion over responsibility and breakdowns in the claims management and cost reimbursement process. Consequently, we are encouraged that NPFC intends to update its procedures to clarify responsible party designations.

End Notes

[1] Pub. L. No. 101-380, 104 Stat. 489 (1990).

[2] The Fund also pays for the costs of certain federal agency operations.

[3] BP established the Trust under Delaware law, which generally provides that the principal of the trust can be used only for the purposes stated in the trust agreement and that the terms of the trust agreement cannot be modified and are legally enforceable by the trustees. BP pledged collateral to cover its funding commitment to the Trust.

[4] The Jones Act 46 U.S.C. § 30104, establishes liability for injury or death of seamen incurred in the course of their employment.

[5] GAO, *Deepwater Horizon Oil Spill: Preliminary Assessment of Federal Financial Risks and Cost Reimbursement and Notification Policies and Procedures*, GAO-11-90R (Washington, D.C.: Nov. 12, 2010). [6]The $1-billion cap is concurrent with a $500-million cap on expenditures for natural resource damages and related assessments.

End Notes for Enclosure

[1] GAO, *Deepwater Horizon Oil Spill: Preliminary Assessment of Federal Financial Risks and Cost Reimbursement and Notification Policies and Procedures*, GAO-11-90R (Washington, D.C.: Nov. 12, 2010).

[2] Pub. L. No. 101-380, 104 Stat. 489 (1990).

[3] On May 11, 2010, NPFC notified BP and Transocean Holdings Incorporated that BP's advertising and claims processing were sufficient, and Transocean should not advertise and should coordinate claims processing with BP. According to NPFC officials, NPFC wanted to avoid public confusion and have only one responsible party advertise for claims.

[4] The three trustees are Citigroup Trust-Delaware, N.A., which serves as the corporate trustee, and John S. Martin, Jr. and Kent D. Syverud, who serve as individual trustees.

[5] The funding schedule for the escrow account agreed to by the administration and BP was for contributions by BP of $5 billion a year for 4 years. BP later confirmed that the funding schedule would include an initial deposit of $3 billion, which was made on August 9, 2010, with an additional deposit of $2 billion made in the fourth quarter of 2010 and $1.25 billion a quarter thereafter until the entire $20 billion has been deposited.

[6] The Jones Act 46 U.S.C. § 30104, establishes liability for injury or death of seamen incurred in the course of their employment.

[7] The $1-billion cap is concurrent with a $500-million cap on expenditures for natural resource damages and related assessments.

[8] An obligation is a commitment, such as a contract, that creates a legal liability for the payment of goods and services ordered or received. NPFC's procedures for monitoring the amount spent toward the cap use the actual expenditures and obligated amounts.

[9] Federal Authorizations authorize reimbursement of federal and nonfederal government agencies from the Fund for oil-spill-response and removal activities. NPFC uses MIPRs rather than Federal Authorizations for the Department of Defense and certain other agencies.

[10] According to the Coast Guard, direct costs are operating costs that it otherwise would not have incurred but for the oil spill.

[11] There is a statutory cap of $500 million in expenditures from the Fund per incident for natural resource damage assessments and claims. 26 U.S.C. § 9509(c)(2).

[12] According to GCCF's Web site, claims under Emergency Advanced Payments could be submitted through November 23, 2010, and were adjudicated and paid through December 15, 2010. Since the deadline expired, this payment is no longer available.

[13] Both Quick Pay and Full Review Claims require claimants to sign a release waiving any rights they may have against responsible parties, to file or participate in legal action, or to submit any claim to NPFC for payment.

[14] Although the contract is currently effective through October 31, 2011, NPFC officials said they could issue additional task orders, so that the contract could be used to provide NPFC with additional support operations through October 31, 2012.

[15] Under OPA and the implementing federal regulations and policies, if the designated responsible party denies a claim or does not settle it within 90 days, a claimant may commence action in court against the responsible party or present a claim to NPFC.

[16] *Claimant Review Status as of March 31, 2011*, http://www.gulfcoastclaimsfacility.com (accessed April 1, 2011).

[17] Numerous individuals, businesses, states, and the federal government have commenced various actions in a number of courts against several companies, including BP, seeking damages or declaratory or injunctive relief under several laws, including OPA. Many of these pending

cases have been consolidated in multidistrict litigation in the U.S. District Court for the Eastern District of Louisiana. See http://www.laed.uscourts.gov/OilSpill/OilSpill.htm.

[18] If a claimant decides to commence litigation against the responsible parties, NPFC will not review the same claim until the litigation has concluded.

[19] The Omnibus Budget Reconciliation Act of 1986, Pub. L. No. 99-509, established the Fund and its original expenditure caps, and OPA modified the expenditure caps to their current level. *See* 26 U.S.C. § 9509(c)(2).

In Deepwater Horizon: ISBN: 978-1- 61942-499-9
Editors: D. A. Michaels and G. S. Aston © 2012 Nova Science Publishers, Inc

Chapter 3

DEEPWATER HORIZON OIL SPILL: PRELIMINARY ASSESSMENT OF FEDERAL FINANCIAL RISKS AND COST REIMBURSEMENT AND NOTIFICATION POLICIES AND PROCEDURES*

United States Government Accountability Office

Washington, DC 20548

November 12, 2010
Congressional Requesters

On April 20, 2010, an oil spill of national significance in the Gulf of Mexico followed an explosion on the mobile offshore drilling unit *Deepwater Horizon* (the *Deepwater Horizon* oil spill). The *Deepwater Horizon* was leased by BP America Production Company (BP) as part of the Macondo project.[1] 152 days later, on September 19, 2010, BP confirmed the completion of cementing operations to prevent further oil from spilling from the Macondo Prospect well to which the *Deepwater Horizon* was attached when it exploded.

---

* This is an edited, reformatted and augmented version of the United States Government Accountability Office publication, GAO-11-90R Oil Spill Liability Trust Fund, dated November 12, 2010.

In order to coordinate the federal response to the *Deepwater Horizon* oil spill, the National Incident Commander established the Deepwater Integrated Services Team (IST) consisting of 18 federal agencies, including the U.S. Coast Guard and the Department of Justice (DOJ).

The U.S. Coast Guard's National Pollution Funds Center (NPFC) designated two BP subsidiaries—BP Exploration and Production and its guarantor, BP Corporation North America, Inc.—and five other companies as "Responsible Parties" for *Deepwater Horizon* oil spill related claims. Shortly after the spill, at the direction of NPFC, BP established a facility to receive and process all claims against Responsible Parties. In June 2010, as part of an oral agreement between the administration and BP, BP established a new claims processing facility—the Gulf Coast Claims Facility (GCCF). GCCF began operations on August 23, 2010, and is responsible for handling claims from individuals and businesses for damages resulting from the *Deepwater Horizon* oil spill. BP also established an irrevocable trust (Trust), to which BP is to provide a total of $20 billion by 2014, primarily for the purpose of paying GCCF and other claims related to the *Deepwater Horizon* oil spill.[2] (Enclosure IV provides a timeline of *Deepwater Horizon* oil spill events.)

The total cost to clean up this massive and potentially unprecedented spill, the damage to the environment, as well as the potential impact to the livelihood and economic status of businesses and individuals in the region will undoubtedly be significant, with current estimates from BP and Oxford Economics in the tens of billions of dollars.[3] However, the full extent of such costs and the extent to which they will ultimately be paid by the Responsible Parties or federal, state, and local governments is unknown at this time and depends on a variety of factors.

The complex legal framework in place for oil spill liability and response funding will play an integral role in determining who is responsible and will ultimately pay the costs associated with the *Deepwater Horizon* oil spill. In this regard, the Oil Pollution Act of 1990, [4] as amended (OPA), which Congress enacted after the *Exxon Valdez* spill in 1989, authorized use of the Oil Spill Liability Trust Fund (Fund), which is administered by NPFC and is subject to certain caps on the amount of its expenditures. The Fund was established to pay for certain oil spill cleanup costs and damages using federal tax revenues for immediate response costs and when the Responsible Parties cannot be identified or do not pay. OPA also provided that the federal government may subsequently seek reimbursement for these costs from Responsible Parties.[5] Since the *Deepwater Horizon* oil spill, a number of related legislative proposals have been introduced. For example, in May 2010, the administration made several proposals including increasing the caps on payments from the Fund and obtaining express authority to oversee Responsible Parties' claims processing.

The objectives of this study were to provide a preliminary assessment of (1) financial risks and exposures to the Fund and the federal government as a result of the *Deepwater Horizon* oil spill, and (2) Coast Guard's NPFC cost reimbursement policies and procedures for *Deepwater Horizon* oil spill costs. In addition, we provide a description of the framework for federal monitoring and oversight efforts over the Responsible Parties for the *Deepwater Horizon* oil spill, including federal efforts to oversee BP's and GCCF's *Deepwater Horizon* oil spill claims payments. This product is the first of a planned body of work to evaluate and assess the federal risks and exposures resulting from the *Deepwater Horizon* oil spill. Our follow on work is to include assessing and testing NPFC Fund reimbursement policies and procedures and analyzing the framework for federal monitoring and oversight efforts over the Responsible Parties for the *Deepwater Horizon* oil spill.

Scope and Methodology

To identify and provide a preliminary assessment of the financial risks to the Fund and the federal government as a result of the *Deepwater Horizon* oil spill, we identified and analyzed applicable laws and regulations and reviewed our previous work on the cost of oil spills to identify statutory and regulatory limitations on the liability of Responsible Parties that may pose financial risks to the Fund and federal government to the extent that *Deepwater Horizon* oil spill claims exceed such limits.[6]

To provide a preliminary assessment of the design of NPFC's reimbursement process, we interviewed NPFC officials, performed walkthroughs, and conducted a review of NPFC's documented policies and procedures. We also obtained invoices NPFC sent to the Responsible Parties to reimburse the Fund and analyzed Fund billings from federal and state agencies. Our assessment was preliminary in that we did not test the reimbursement invoices to determine their accuracy or eligibility under OPA.

To provide an overview of the framework for federal monitoring and oversight of BP's claims payment process, we interviewed DOJ officials and other cognizant federal and state officials regarding coordination of federal government agencies efforts to monitor and oversee BP's claims process and payment activities. We did not evaluate the effectiveness of federal monitoring and oversight efforts over BP's claims process and the GCCF. (Enclosure II provides further details of our scope and methodology.)

On November 9, 2010, we briefed your staff on the results of our work. This letter summarizes the information provided during the briefing. The briefing slides are provided in enclosure I.

We conducted this performance audit from July 2010 to October 2010, in accordance with generally accepted government auditing standards. Those standards require that we plan and perform the audit to obtain sufficient, appropriate evidence to provide a reasonable basis for our findings and conclusions based on our audit objectives. We believe that the evidence obtained provides a reasonable basis for our findings and conclusions based on our audit objectives.

SUMMARY OF RESULTS

Because the total costs of the *Deepwater Horizon* oil spill are still unknown, the federal government's financial exposure as a result of the oil spill is also unknown. Additionally, the fiscal exposure of the federal government resulting from the *Deepwater Horizon* oil spill could increase as Responsible Parties' capacity to cover oil spill-related costs is reduced whether as a result of extensive oil spill liabilities, poor financial condition, or other factors. BP has voluntarily established a Trust to be funded incrementally up to $20 billion, has paid other costs outside of the Trust, and has stated that it will continue to pay additional costs. BP's financial condition and its continuing resolve to stand behind its public commitments will be key factors if additional costs need to be paid.

Certain statutory limits on the amount of federal funds available for response costs and damages are intended to mitigate the exposure. For example, OPA establishes caps on the amount of funds that can be expended on each oil spill. (Enclosure III discusses the statutory liability and response framework associated with oil spills.)

NPFC has billed the Responsible Parties for the *Deepwater Horizon* oil spill $581 million for response activities performed by nine federal government agencies and various state government agencies. After NPFC authorizes reimbursement, the government agencies are paid from the Fund for actual expenditures. BP has paid NPFC $518.4 million as of October 12, 2010. The Fund is at risk of reaching the OPA-established $1 billion per incident cap on total expenditures in the relatively near future. [7] Consequently, unless the statute is amended to exclude amounts reimbursed by Responsible Parties from the cap, the Fund may be unable to pay any OPAcompensable claims or other *Deepwater Horizon* oil spill-related costs above that limit.[8]

Our preliminary assessment of the design of Coast Guard's NPFC's policies and procedures for obtaining reimbursement for *Deepwater Horizon* oil spill costs found they did not always reflect current practices and were not sufficiently detailed to ensure they could be followed consistently. For example, NPFC's procedures for identifying and notifying Responsible Parties are dated 1996, when the Coast Guard was part of Department of Transportation, and are marked "draft." Also, in responding to the ongoing nature of the costs associated with the *Deepwater Horizon* oil spill, NPFC changed its practice of billing Responsible Parties from solely billing for federal agencies' expenditures to also including an amount reflecting a percentage of federal agencies' obligations.[9] Although it began using this new billing practice in June 2010, as of October 4, 2010, NPFC had not yet updated its policies and procedures manual to reflect this new practice. Further, the Department of Defense (DOD) has been using Military Interdepartmental Purchase Requests (MIPR) to provide NPFC with its requests for cost reimbursement since 1990. As a matter of practice, NPFC has been accepting and processing DOD requests for *Deepwater Horizon* oil spill cost reimbursements using MIPRs. However, NPFC has not documented its practices for these requests in its current policies and procedures manual, which would help ensure these practices are fully and consistently followed. NPFC is at risk of these procedures being out of date and not reflecting management's current directives. In addition, existing procedures for implementing NPFC's policy requiring formal notification to Responsible Parties were not consistent. Consequently, we found that NPFC procedures followed in notifying the *Deepwater Horizon* oil spill Responsible Parties were not always effective. For example, a *Deepwater Horizon* oil spill Responsible Party stated during a July 2010 hearing that his company had not received notification of designation.

The federal government has been involved in overseeing Responsible Parties' claims processing resulting from the *Deepwater Horizon* oil spill. Following the spill, DOJ, the Department of Homeland Security (DHS) and various other federal agencies have been overseeing the establishment of a claims process and monitoring claims processing activities by BP on behalf of the designated Responsible Parties. For example, DOJ encouraged BP to set aside substantial resources to protect federal interests related to *Deepwater Horizon* claims payment. Further, in September 2010, DOJ notified GCCF's administrator of its dissatisfaction regarding GCCF's pace of processing *Deepwater Horizon* oil spill claims, and urged GCCF to devote additional resources (or make administrative changes as necessary) to meet its own claims processing timeliness standards. In addition, DOJ is helping to coordinate investigations of fraudulent claims associated with the *Deepwater Horizon* oil spill. The DHS, through its component agencies—the Coast Guard and the Federal Emergency Management Agency—was also engaged in various *Deepwater Horizon* oil spill-related claims processing oversight and monitoring activities. For example, the Coast Guard, in its capacity as the Federal On-Scene Coordinator for the *Deepwater Horizon* oil spill verified whether costs incurred for removal were consistent with the National Contingency Plan (NCP).[10]

In addition to DHS, over 15 federal agencies have been involved in overseeing Responsible Parties' efforts to address the effects of the *Deepwater Horizon* oil spill through their participation in the Deepwater IST. The Deepwater IST was established to, among other things, coordinate the federal government's oversight of BP's claims processing to avoid duplicate payments. However, as of September 30, 2010, the Deepwater IST was to transfer ongoing oversight to appropriate federal agencies or workgroups.

CONCLUSION

Financial risks exist for the federal government as a result of the *Deepwater Horizon* oil spill. Total costs associated with the oil spill are unknown, but estimates are in the tens of billions of dollars. BP has committed to paying costs from the *Deepwater Horizon* oil spill, even to the extent that such costs exceed the $20 billion it has agreed to set aside for such costs. OPA places limits on the liability of the Responsible Parties, but BP has stated that it will continue to pay costs above the limits applicable to the *Deepwater Horizon* oil spill. However, financial risks to the federal government exist if circumstances occur that adversely impact BP or other Responsible Parties' financial condition or ability to pay such claims. As such, it remains unclear whether the federal government will ultimately have to cover *Deepwater Horizon* oil spill-related costs, and if so, the potential amounts that could be involved.

Consequently, in order to help minimize federal financial risks associated with the *Deepwater Horizon* oil spill, it will be imperative that the federal government take prompt action to ensure that its policies and procedures are up-to-date, clear, and sufficiently detailed. Further, going forward, sustained federal oversight and monitoring—including establishing mechanisms to ensure that policies and procedures are fully and effectively implemented—and enacting legislation concerning the $1 billion per incident cap will be essential.

MATTER FOR CONGRESSIONAL CONSIDERATION

Congress should consider amending OPA or enacting new legislation that eliminates the Fund's $1 billion per incident expenditure cap to the extent that it does not take into account reimbursements from Responsible Parties. In this regard, Congress may want to consider setting a Fund cap associated with an incident based upon net expenditures (expenditures less reimbursements).

RECOMMENDATIONS FOR EXECUTIVE ACTION

In order to help establish and maintain effective cost reimbursement policies and procedures for the Fund, we recommend that the Secretary of Homeland Security direct the Director of the U.S. Coast Guard's NPFC to update NPFC's policies and procedures to include

- current Fund reimbursement billing practices that reflect both a percentage of federal agencies' obligations as well as expenditures, and
- specific procedural guidance on processing DOD requests for reimbursement using MIPRs.

In order to ensure that all Responsible Parties are properly notified of their responsibilities for an oil spill, we recommend that the Secretary of Homeland Security direct the Director of NPFC to

- update NPFC's current policies to reflect current organization and structure and managements' directives, and
- update NPFC's current procedures to provide detailed guidance and procedures for identifying and documenting all Responsible Party notifications.

Agency Comments

We provided a draft of our slides to DHS, the Department of the Interior, and DOJ's management for comment. We received written comments from the Special Assistant, Office of the Chief Financial Officer, DHS, which are contained in enclosure V. We also received a subsequent email from DHS providing additional comments related to GAO's four recommendations. DOJ provided technical comments which we considered and incorporated as appropriate. The Department of the Interior did not provide any comments.

In its written comments, DHS agreed with our matter for congressional consideration of revising the current $1 billion per incident cap and cited a May 2010 request to increase the per incident cap to $1.5 billion. With respect to our recommendations, DHS stated that while its existing policies and procedures are sound to address many oil spills, the significance of the *Deepwater Horizon* spill highlights the need for adopting policy refinements. Accordingly, DHS agreed to update NPFC's current policies to reflect current organization and structure and stated it will further evaluate and analyze our recommendation to update NPFC's Fund reimbursement billing policies and procedures.

However, DHS commented that no additional actions were needed with respect to two of our recommendations. We disagree. For our recommendation to provide specific procedural guidance on processing DOD requests for reimbursements using MIPRs, DHS stated that existing legally binding bilateral agreements with documented procedures are already in place. However, NPFC's *User Reference Guide* does not address procedures for processing MIPRs, nor does it provide specific examples on how to complete a MIPR or provide guidance on when a MIPR should be used. With respect to our recommendation to provide detailed guidance and procedures for identifying and documenting all Responsible Parties notifications, DHS viewed our recommendation as being directed toward using such formal notifications in determining Responsible Parties' liability. While we recognize that the purpose of the notice of designation is to help ensure Responsible Parties advertise the claims process as quickly as possible, we believe that failure to provide prompt formal notifications may result in confusion, hindering Responsible Parties' compliance with cost recovery processes. Consequently, we continue to believe that actions are needed to implement all four of our recommendations.

DHS's comments also included a number of technical suggestions concerning our findings which we considered and incorporated as appropriate in finalizing our product. For example, we added clarification concerning DHS actions to deliver designation notices to all potential Responsible Parties.

Susan Ragland
Director
Financial Management and Assurance

List of Requesters

The Honorable John Conyers, Jr.
Chairman
Committee on the Judiciary
House of Representatives

The Honorable Nick J. Rahall, II
Chairman
Committee on Natural Resources
House of Representatives

The Honorable Tom Carper
Chairman
Subcommittee on Federal Financial Management,
Government Information, Federal Services,
and International Security
Committee on Homeland Security and Governmental Affairs
United States Senate

The Honorable Sheldon Whitehouse
Chairman
Subcommittee on Oversight
Committee on Environment and Public Works United States Senate

The Honorable Michael C. Burgess
Ranking Member
Subcommittee on Oversight and Investigations
Committee on Energy and Commerce
House of Representatives

ENCLOSURE I

Overview

- Background
- Objectives
- Scope and Methodology
- Results
- The Fund and Federal Government Face a Variety of Financial Risks as a Result of the *Deepwater Horizon* Oil Spill
- Federal Policies and Procedures Not Current, Clear, and Sufficiently Detailed
- Federal Government Efforts to Monitor and Oversee Responsible Parties' Claims Processing
- Conclusions
- Matter for Congressional Consideration
- Recommendations for Executive Action
- Agency Comments and Our Evaluation

Background

- On April 20, 2010, an oil spill of national significance followed an explosion on the *Deepwater Horizon* mobile offshore drilling unit located in the Gulf of Mexico.[11] Cementing operations to prevent further oil from spilling from the Macondo Prospect well was completed on September 19, 2010. (See enclosure IV for a timeline of *Deepwater Horizon* oil spill events.)
- A complex landscape of laws and regulations governs the liability for oil spill costs of different parties.
- Injuries and damages that arise from an oil spill incident are governed by federal statutes and common law, federal securities laws, and various state laws.
- For example, the Oil Pollution Act of 1990 (OPA), as amended, [12] places the primary liability for the cost of the oil spills—up to certain limits—on the Responsible Party or Parties and establishes the limits on liability of Responsible Parties for removal costs and damages specified in OPA (referred to as OPA compensable damages).[13]

- The U.S. Coast Guard's National Pollution Funds Center (NPFC) administers the Oil Spill Liability Trust Fund (Fund) by disbursing funds for federal cleanup and natural resource restoration, monitoring the sources and uses of the Funds, adjudicating claims submitted to the Fund for payment, and pursuing reimbursements from Responsible Parties for costs and damages paid by the Fund and certain other recoverable costs.
- Under OPA, the authorized limit on expenditures to be paid by the Fund is currently $1 billion in total expenditures per incident, with a concurrent limit of $500 million per incident for natural resource damage assessments and claims. According to the Department of Homeland Security, federal agencies may separately incur costs and fund them with agency appropriations, if available for that purpose.

Background – Oil Spill Liability Trust Fund

- The Fund's current revenue sources include
- a per barrel tax of 8 cents on petroleum products either produced in the United States or imported from other countries,
- recoveries from Responsible Parties for costs of removal and damages,
- fines and penalties paid pursuant to various statutes, and
- interest earned on U.S. Treasury investments.
- As shown in figure 1, the Fund's balance has varied over the years.
- For fiscal years 2003 through 2007, its balance was less than the $1 billion limit on expenditures for the response to a single spill. This was in part because the Fund's per barrel tax on U.S. produced and imported oil was not collected for most of 1995 through 2006, because the barrel tax had expired in December 1994. The Energy Policy Act of 2005 reinstituted the 5 cents per barrel tax on produced and imported oil in April 2006, which the Energy Improvement and Extension Act of 2008 subsequently increased to 8 cents per barrel.[14] On September 30, 2010, the unaudited Fund balance was $1.69 billion.

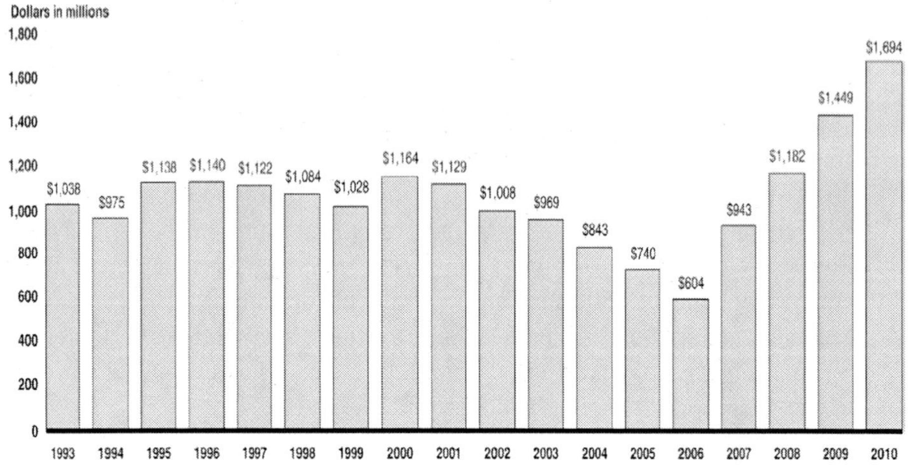

Source: GAO analysis of NPFC data.

Note: The Fund balance increase in 2000 was largely because of a transfer of $181.8 million from the Trans-Alaska Pipeline Liability Fund.

Figure 1. Oil Spill Liability Trust Fund Balance, 1993–2010 (Unaudited).

Background – Federal Agencies' Roles

- Responding to oil spills involves a coordinated effort by various parties, including (1) Coast Guard or Environmental Protection Agency (EPA) as the Federal On-Scene Coordinator;[15] (2) federal, state, local, and Indian tribal government agencies; (3) private companies that specialize in oil spill clean up;[16] and (4) the Responsible Parties, their guarantors (i.e., insurers), and qualified individuals designated by responsible parties to respond to oil spills.
- To fund government agencies' oil spill removal costs, the Federal On-Scene Coordinators
- issue Pollution Removal Funding Authorizations (federal authorizations) or Military Interdepartmental Purchase Requests (MIPR)[17] to quickly obtain services and assistance from government agencies,
- verify that the services or goods were received and consistent with the National Contingency Plan and certify the supporting cost documentation, and
- send the cost documentation to NPFC, which authorizes the Coast Guard's Finance Center to pay the government agencies.

Background –Trust and Gulf Coast Claims Facility

- The Coast Guard, without in any way relieving other Responsible Parties of liability, directed BP to establish a single claims facility for all Responsible Parties to centralize claims processing for claimants.[18]
- On June 16, 2010, BP announced that it would create a $20 billion escrow account to satisfy claims resolved by the Gulf Coast Claims Facility (GCCF) and certain other claims, including natural resource damages.
- BP established an irrevocable Trust (for the announced escrow account) on August 6, 2010, designating three trustees[19] with fiduciary responsibility to collect promised contributions from BP and make disbursements to permitted categories of beneficiaries. It committed BP to fund the Trust on a quarterly basis over 3-1/2 years for a total of $20 billion to be paid into the Trust as of 2014 (as shown in figure 2).[20] The Trust is to pay some OPA-compensable claims and some other claims for personal injuries that are not OPA-compensable, but for which BP would be liable under other federal or state laws, such as the Jones Act or state oil pollution acts.[21]
- BP established the GCCF to provide a mechanism for individuals and businesses to file claims for costs and damages incurred as a result of the *Deepwater Horizon* oil spill. The GCCF began operations and started accepting claim forms on August 23, 2010. The GCCF, administered by Kenneth R. Feinberg, draws funds from the Trust to pay claims.
-
-

Background – Gulf Coast Claims Facility

- The GCCF Administrator has established the following principles for GCCF's operations:
- GCCF officials are to evaluate all claims in a prompt and fair manner guided by applicable law,
- The establishment of the GCCF does not diminish any right of any individual or business that existed prior to the GCCF's creation, and
- The GCCF claims process has been structured to comply with OPA and apply the standards of OPA. In addition, it will consider some non-

OPA-compensable claims, such as those for physical injuries, that are based on applicable federal or state laws.

- As of August 2010, GCCF has established the following types of payments:
- Emergency Advance Payments. Payments available to individuals and businesses that are experiencing financial hardship resulting from damages incurred from the *Deepwater Horizon* oil spill. According to GCCF Emergency Advance Payment protocols, claims for Emergency Advance Payments can be submitted through November 23, 2010.
- Final Payments. Payments, over and above any emergency advance payments, to individuals and businesses for total losses incurred because of the *Deepwater Horizon* oil spill. According to the Frequently Asked Questions on GCCF's Web site,[22] the GCCF will be accepting claims for Final Payments that are submitted through August 23, 2013.
- These payments are intended to provide compensation for both OPA-compensable and non-OPAcompensable claims.

Background – The Trust

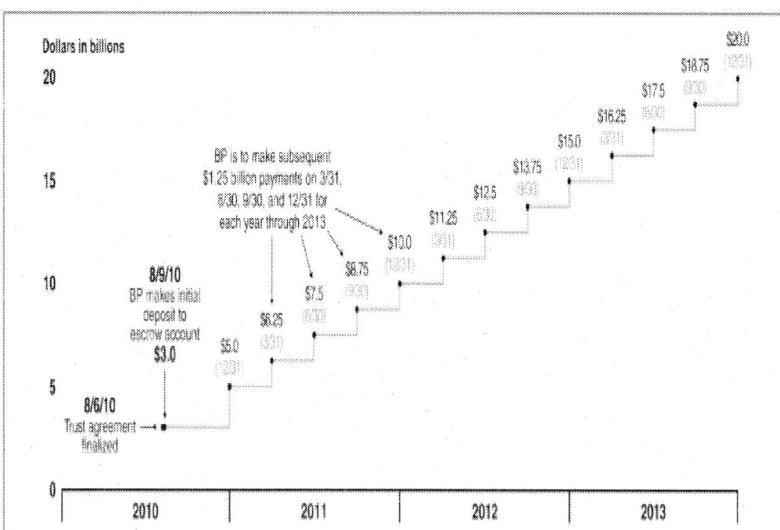

Source: GAO analysis on an August 9, 2010, BP press release on the funding of the $20 billiom account.

Figure 2. The Trust's $20 Billion Funding Time Frame.

Objectives

- Our objectives for this briefing are to
- provide a preliminary assessment of financial risks to the Fund and the federal government as a result of the *Deepwater Horizon* oil spill;
- provide a preliminary assessment of the Coast Guard's NPFC cost reimbursement policies and procedures for *Deepwater Horizon* oil spill costs; and
- describe the framework for federal monitoring and oversight efforts over the Responsible Parties for the *Deepwater Horizon* oil spill, including federal efforts to oversee BP's and GCCF's *Deepwater Horizon* oil spill claims payments.

Scope and Methodology

- To address these objectives, we
- identified and analyzed applicable laws and regulations and our previous work[23] to identify statutory and regulatory limitations on the liability of Responsible Parties that may pose risks to the Fund and federal government;
- interviewed Fund officials, performed walkthroughs, reviewed and summarized invoices sent to the Responsible Parties for reimbursement, and reviewed relevant policies and procedures to make a preliminary assessment of the Fund's cost reimbursement policies and procedures; and
- interviewed Department of Justice (DOJ) and Deepwater Integrated Services Team (IST) officials and reviewed relevant coordination and status updates to obtain an understanding of federal monitoring and oversight of BP's claims payment activities.
- This product is the first of a planned body of work to evaluate and assess the federal risks and exposures resulting from the *Deepwater Horizon* oil spill. Our follow on work is to include assessing and testing NPFC Fund reimbursement policies and procedures and analyzing the framework for federal monitoring and oversight efforts over the Responsible Parties for the *Deepwater Horizon* oil spill.
- We conducted this performance audit from July 2010 to October 2010, in accordance with generally accepted government auditing standards. Those standards require that we plan and perform the audit to obtain sufficient, appropriate evidence to provide a reasonable basis for

our findings and conclusions based on our audit objectives. We believe that the evidence obtained provides a reasonable basis for our findings and conclusions based on our audit objectives.

The Fund and Federal Government Face a Variety of Financial Risks as a Result of the *Deepwater Horizon* Oil Spill

- According to BP's second quarter 2010 projections, the *Deepwater Horizon* oil spill is projected to result in tens of billions of dollars in cleanup, restoration, loss of income, property damage, and other costs.[24]
- BP stated publicly and in filings in federal district court that it will pay all costs related to the *Deepwater Horizon* oil spill, even if costs exceed the statutory limit on its liability for costs of oil spilled from the Macondo Prospect Well (removal costs plus $75 million).[25]
- As of October 15, 2010, NPFC had sent seven bills to the *Deepwater Horizon* Responsible Parties for a cumulative total of $581 million. These bills are primarily to reimburse federal government costs, but also include some other costs, such as the National Guard. BP has stated its intent to recover some of these costs from the other Responsible Parties under the provisions of OPA or applicable contractual agreements.
- NPFC officials told us that they expect the Fund to reach the $1 billion per incident cap at some point for the *Deepwater Horizon* oil spill, but the timeframe for this is uncertain.[26]
- Federal agencies incurring removal and certain other costs related to the *Deepwater Horizon* oil spill face risks as costs increase. If the agencies incur OPA-compensable costs that collectively exceed the $1 billion per incident cap on total expenditures from the Fund, the statute would need to be amended to exclude from the cap amounts recovered from the Responsible Parties in order to obtain reimbursements from the Fund for their costs incurred.[27] Additionally, federal agencies may need to use their annual appropriations or other agency budgetary resources to cover non-reimbursable costs.
- The $581 million billed to the Responsible Parties includes $323.3 million for Pollution Removal Funding Authorizations (Federal Authorizations) and MIPRs and $44.0 million for Coast Guard direct costs, such as those for purchases, contractors, and travel. The amounts billed also include $213.7 million for Coast Guard recoverable, or indirect, costs, such as personnel and equipment.

- The $581 million billed to the Responsible Parties does not necessarily reflect expenditures from the Fund.
- For example, although NPFC bills the Responsible Parties for 75 percent of the Federal Authorization and MIPR ceilings, agencies are reimbursed by the Fund for actual expenditures incurred. The agencies submit cost reimbursement packages with costs incurred to NPFC for approval for reimbursement from the Fund. Not all of the agencies that have incurred costs have submitted cost reimbursement packages to NPFC. According to information provided by NPFC, as of October 19, 2010, NPFC has authorized for reimbursement, approximately $125.5 million, or 38.8 percent of the amount billed to the Responsible Parties for Federal Authorizations and MIPRs.
- Additionally, the Coast Guard recoverable costs are funded through the agency's annual operating expenses appropriation, and the Coast Guard has not sought reimbursement for these costs from the Fund.[28] Therefore, the $213.7 million is not applied to the $1 billion statutory limit, although the Fund benefits from reimbursements from Responsible Parties.[29]
- The $581 million does not include any amounts for natural resource damage assessment initiation costs, although NPFC has an interagency agreement with the Department of the Interior for about $22.4 million for the initiation assessment.[30]
- Further, since the total cost of the *Deepwater Horizon* oil spill has yet to be determined, the federal government remains at risk of having to pay relevant costs above the $20 billion planned to fund the Trust if circumstances change. BP's financial condition and its continuing resolve to stand behind its public commitments will be key factors if additional costs need to be paid.
- BP faces a large number of lawsuits, which potentially could result in final judgments or settlements to be paid from the Trust.
- As of October 21, 2010, according to BP's Web site,[31] it paid a total of $3,198.5 million to reimburse costs incurred by federal and state governments and pay claims to individuals and businesses associated with the *Deepwater Horizon* oil spill.[32] (See table 1.)
- The extent of natural resource damages has yet to be determined, which could result in significant additional costs. To provide a frame of reference, the *Exxon Valdez*'s natural resource damages were at least $1 billion. The Fund has a $500 million statutory limit on natural resource damage assessments and claims.

Table 1. Claims and Government Payments as of October 21, 2010 (Dollars in Millions)

Type	BP direct	Trust	Total
Individuals and businesses	$395.6	$1,481.8	$1,877.4
Fund for real estate brokers and agents		$34.5	$34.5
Federal government – response and removal	$518.4		$518.4
Federal government - other	$63.4		$63.4
States and other	$564.7	$140.1	$704.8
Total Paid	$1,542.1	$1,656.4	$3,198.5

Source: Unaudited data from BP's Web site.

FEDERAL POLICIES AND PROCEDURES NOT CURRENT, CLEAR AND SUFFICIENTLY DETAILED

- Our preliminary assessment of the design of NPFC policies and procedures over the billings to Responsible Parties noted instances where NPFC had not updated its documented policies and procedures to reflect current NPFC practices in the following areas:
- its practices for billing the Responsible Parties for this incident,
- procedures for identifying and notifying Responsible Parties (its *Technical Operating Procedures for Designation of Source*), and
- the specific steps required for the proper processing of MIPRs, which NPFC has been using since 1990 to reimburse the Department of Defense (DOD) for costs incurred with spills and that NPFC stated it has refined as a result of the *Deepwater Horizon* oil spill.

- Because of the ongoing nature of the *Deepwater Horizon* oil spill, NPFC determined that it was necessary to change its practices for billing the Responsible Parties for this incident.
- NPFC developed and implemented a new practice for billing the Responsible Parties for the costs incurred by the Fund. However, NPFC has not updated its policies and procedures to reflect these new practices.
- Prior to the *Deepwater Horizon* oil spill, NPFC's policy was to bill Responsible Parties for actual expenditures associated with a spill.
- NPFC developed a new practice for billing the *Deepwater Horizon* oil spill Responsible Parties based on a combination of obligations and actual expenditures.[33]

- As of October 2010, NPFC had not yet incorporated its new billing practices in its *Case Management Standard Operating Procedures* or other policy manuals.[34]

- NPFC's documented procedures for identifying and notifying Responsible Parties (i.e., its *Technical Operating Procedures for Designation of Source*) are dated 1996 and marked "draft."
- NPFC's *Technical Operating Procedures* were drafted in 1996 when the Coast Guard was part of the Department of Transportation. Since the procedures were drafted, multiple organizational changes have occurred including Coast Guard becoming part of the Department of Homeland Security in 2003 and the establishment of the Department of Interior's Bureau of Ocean Energy Management, Regulation and Enforcement in June 2010.[35]
- Since NPFC's *Technical Operating Procedures* were drafted in 1996, it is important that NPFC update them to ensure that they reflect the current organization and structure as well as management's current directives.

- NPFC did not follow its *Technical Operating Procedures* in notifying the *Deepwater Horizon* oil spill Responsible Parties.
- On April 28, 2010, NPFC sent formal Notices of Designation (advising the companies of BP's and Transocean's[36] obligation to advertise for claims for the *Deepwater Horizon* oil spill) to (1) BP Exploration & Production; (BP), (2) BP's guarantor, BP Corporation North America, Inc.; (3) Transocean Holdings Incorporated (Transocean); and (4) Transocean's guarantor, QBE Underwriting, LTD.
- However, NPFC officials told us that in the case of the *Deepwater Horizon* oil spill, as of October 2010, NPFC had identified seven Responsible Parties, but had notified three of the Responsible Parties identified—(1) Anadarko E&P Company, LP, (2) Anadarko Petroleum Corporation and (3) MOEX Offshore 2007 LLC—only through invoices sent to recover costs.
- NPFC *Technical Operating Procedures* provide that Responsible Parties and their guarantors are to be identified and formally notified of their oil spill–related responsibilities through a Notice of Designation.[37] The Notice of Designation states that if the Responsible Party does not deny the designation, then the Responsible Party is required to advertise the procedures by which persons may submit claims.

- Because multiple parties were involved in the *Deepwater Horizon* oil spill, it is important that NPFC appropriately identify, document, and notify the Responsible Parties involved to ensure that the appropriate Responsible Parties are held responsible for advertising.
- NPFC officials view the Responsible Party designation process as applicable only to the assignment of claims advertising responsibilities under OPA and not as a determination of liability, and the liability of any particular entity will be addressed through investigation and possible litigation. However, NPFC policies and procedures require an assessment of liability in its designation process and NPFC uses responses to its designations in its cost reimbursement process. The failure to follow documented notification procedures creates risks of confusion and breakdowns in the claims management and cost reimbursement process.

- NPFC's current policy and procedures manual does not address the specific steps required for the proper processing of MIPRs, which NPFC has been using since 1990 to reimburse DOD for costs incurred responding to oil spills.
- NPFC's *User Reference Guide*, which contains NPFC's policies and procedures,
- provides specific examples of Federal Authorizations and when Pollution Removal Funding Authorizations should be used,
- does not mention MIPRs, and thus does not provide specific examples on how to complete an MIPR or discuss when an MIPR should be used, and
- provides no information regarding invoicing and reimbursing MIPRs.
- As the result of the *Deepwater Horizon* oil spill, NPFC stated that it has refined its process related to MIPRs and the agency stated that it will be formally documenting and promulgating these refinements.
- Our *Standards for Internal Control in the Federal Government*[38] provide that internal control and all transactions and other significant events need to be clearly documented and readily available for examination. The documentation should appear in management directives, administrative policies, or operating manuals. All documentation and records should be properly managed and maintained.
- Without clearly documented policies and procedures for dealing with the new and current practices established in response to the *Deepwater Horizon* oil spill, NPFC is at risk of its billing practices not being efficient and effective.

Federal Government Efforts to Monitor and Oversee Responsible Parties' Claims Processing

- Federal agencies, including DOJ and the Coast Guard and the Federal Emergency Management Agency (FEMA) within DHS, have participated in monitoring and overseeing the claims and cost reimbursement process for the *Deepwater Horizon* oil spill. In addition, the federal government established the Deepwater IST to lead the government's efforts. The following is additional detail on the roles of the federal agencies:
- DOJ
- Encouraged BP to establish the Trust and the GCCF.[39]
- Reviewed and provided written comments on GCCF's *Protocol for Emergency Advance Payment.*
- Is monitoring GCCF's claims process to determine if it is independent, fair, and efficient.
- Is coordinating investigations of fraudulent claims and has also begun receiving allegations through its National Center for Disaster Fraud.

- Coast Guard

- Acts as the Federal On-Scene Coordinator for the *Deepwater Horizon* oil spill and provides oversight of the response activities.
- Verifies whether a particular cost was incurred for removal and was consistent with the National Contingency Plan.
- NPFC
- Designated Responsible Parties and sent Notices of Designation.
- Generated and sent invoices to Responsible Parties to recover costs.
- Monitors the actions of Responsible Parties to meet their legal obligations.
- Notifies Responsible Parties of claims it receives.

- Deepwater IST (see figure 3)

- Led by DHS's FEMA.
- Was established on June 4, 2010, to coordinate efforts of federal departments and agencies to provide support services and monitor claims in response to the *Deepwater Horizon* oil spill.

• Since established, coordinated interagency and intergovernmental efforts to monitor the status of claims submitted to BP and the effectiveness and efficiency of BP's claims process. Took steps to raise awareness of concerns related to payment policy clarity for claimants, data access and reporting, and coordination of federal/state benefits and services to avoid duplicate payments.

• According to the IST Transition Plan, is to stand down the majority of activities by September 30, 2010, with objectives and priorities either completed or transitioned to the appropriate workgroup or agency for longer-term follow-up.[40]

• As of early September 2010, according to its status report, was expected to have met its general mission by late September.

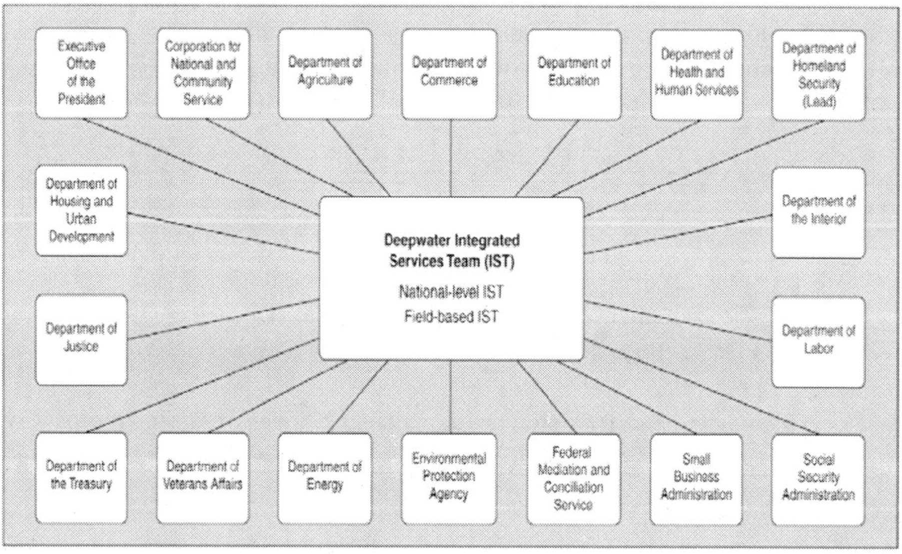

Source: Deepwater IST.

Figure 3. Participants in the Deepwater IST.

CONCLUSION

• Financial risks exist for the federal government as a result of the *Deepwater Horizon* oil spill. Total costs associated with the oil spill are unknown, but estimates are in the tens of billions of dollars. Further, BP has committed to paying costs of the *Deepwater Horizon* oil spill, even to the extent such costs exceed the $20 billion it has agreed to set aside for such costs. However, circumstances may occur that adversely impact BP or other Responsible Parties' financial condition or ability to pay such claims. OPA places conditional limits on the liability of the Responsible Parties, but BP has stated that it will continue to pay costs above the limits applicable to the *Deepwater Horizon* oil spill. Any applicable limits have not been definitively determined through investigation or litigation. As such, it remains unclear whether the federal government will ultimately have to cover any *Deepwater Horizon* oil spill-related costs, and if so, the potential amounts that could be involved.

- Consequently, in order to help minimize federal financial risks associated with the *Deepwater Horizon* oil spill, it will be imperative that the federal government take prompt action to ensure that its policies and procedures are up-to-date, clear, and sufficiently detailed. Further, going forward, sustained federal oversight and monitoring—including establishing mechanisms to ensure that established policies and procedures are fully and effectively implemented—and enacting legislation concerning the $1 billion per incident cap will be essential.

Matter for Congressional Consideration

- Congress should consider amending OPA or enacting new legislation that eliminates the Fund's $1 billion per incident expenditure cap to the extent that it does not take into account reimbursements from Responsible Parties. In this regard, Congress may want to consider setting a Fund cap associated with an incident based upon net expenditures (expenditures less reimbursements).

Recommendations for Executive Action

- In order to help establish and maintain effective cost reimbursement policies and procedures for the Fund, we recommend that the Secretary of Homeland Security direct the Director of the Coast Guard's NPFC to update NPFC's policies and procedures to include
- current Fund reimbursement billing practices that reflect both a percentage of federal agencies' obligations as well as expenditures, and

- specific procedural guidance on processing DOD requests for reimbursement using Military Interdepartmental Purchase Requests.
- In order to ensure that Responsible Parties are properly notified of their responsibilities for an oil spill, we recommend that the Secretary of Homeland Security direct the Director of NPFC to
- update NPFC's current policies to reflect current organization and structure and management's directives, and
- update NPFC's current procedures to provide detailed guidance and procedures for identifying and documenting Responsible Party notifications.

Agency Comments and Our Evaluation

- We provided a draft of our slides to DHS, the Department of Interior, and DOJ's management for comment. We received written comments from the Special Assistant, Office of the Chief Financial Officer, DHS. (See enclosure V.) We also received an email from DHS providing additional comments related to GAO's four recommendations. DHS's comments also included a number of technical suggestions concerning our findings which we considered and incorporated as appropriate in finalizing our product. DOJ also provided technical comments which we considered and incorporated as appropriate. The Department of the Interior did not provide any comments.
- In its written comments, DHS agreed with our matter for congressional consideration of revising the current $1 billion per incident cap and cited a May 2010 request to increase the per incident cap to $1.5 billion.
- With respect to our recommendations, DHS said that while its existing policies and procedures are sound to address many oil spills, the significance of the *Deepwater Horizon* spill highlights the need for adopting policy refinements. Accordingly, DHS agreed to update NPFC's current policies to reflect current organization and structure and stated it will further evaluate and analyze our recommendation to update NPFC's Fund reimbursement billing policies and procedures.

- DHS commented that no additional actions were needed with respect to two of our recommendations. We disagree.
- For our recommendation to provide specific procedural guidance on processing DOD requests for reimbursements using MIPRs, DHS stated that existing legally binding bilateral
- agreements with documented procedures are already in place. However, NPFC's *User Reference Guide* does not address procedures for MIPRs.
- With respect to our recommendation to provide detailed guidance and procedures for identifying and documenting all Responsible Party notifications, DHS viewed our recommendation as being directed toward using such formal notifications in determining Responsible Parties' liability. While we recognize that the purpose of the notice of designation is to help ensure Responsible Parties advertise the claims process as quickly as possible, we believe that failure to provide such prompt formal notifications may result in confusion, hindering Responsible Parties' compliance with cost recovery processes.

ENCLOSURE II: SCOPE AND METHODOLOGY

To identify and analyze financial risks and exposures to the Oil Spill Liability Trust Fund (Fund) and federal government, we identified and analyzed applicable laws and regulations to identify statutory and regulatory limitations on the liability of Responsible Parties that may pose financial risks to the Fund and federal government to the extent that *Deepwater Horizon* oil spill claims exceed such limits. We also considered the risks and exposures identified in our previous work on the cost of oil spills in terms of their potential impact on the Fund.[41]

To provide a preliminary assessment of Coast Guard's National Pollution Funds Center's (NPFC) cost reimbursement process, we interviewed NPFC officials and performed walkthroughs of the cost reimbursement process with NPFC case officers to obtain an understanding of the process. We also observed NPFC's process for creating the fifth invoice that was sent to the Responsible Parties. Additionally, we performed a preliminary review of NPFC's policy and procedure manuals to determine the sufficiency of NPFC's documented policies and procedures. We also reviewed the Department of Homeland Security's (DHS) fiscal year 2009 agency financial report, including the independent auditor's report[42] and DHS Office of Inspector

General's (OIG) *Information Technology Management Letter for the United States Coast Guard Component of the FY 2009 DHS Integrated Audit*[43] to determine the extent to which the independent public auditor and OIG identified weaknesses and deficiencies in the Coast Guard's cost reimbursement or payment processes. We also obtained the invoices that NPFC sent to the Responsible Parties. We checked the invoices for mathematical accuracy and reviewed and summarized the amounts by federal and state agencies; however, we did not test the amounts contained therein to determine their accuracy or eligibility under the Oil Pollution Act (OPA). During our future work, we plan on testing reimbursement invoices to determine their accuracy. We also obtained the Notices of Designation NPFC sent to Responsible Parties and interviewed NPFC officials about their methodology for identifying Responsible Parties and their procedures for notifying them.

To describe federal monitoring and oversight efforts of BP America Production Company (BP) and the Gulf Coast Claims Facility (GCCF) claims process, we interviewed Department of Justice (DOJ) officials about their oversight activities of BP claims process, the establishment of the Trust, and the setup of the GCCF. We also reviewed DOJ's comments on draft GCCF Emergency Advanced Payment protocols and obtained and reviewed the trust agreement. Additionally, we interviewed Deepwater Integrated Services Team (IST) officials about their oversight activities of BP and GCCF claims process and social services coordination efforts. We reviewed the Deepwater IST's documents including its coordination plan, team updates, and transition plan. We did not evaluate the effectiveness of the monitoring and oversight efforts by DOJ and the Deepwater IST. Additionally, we interviewed officials at the Department of the Interior to obtain an understanding of the agency's response to the *Deepwater Horizon* oil spill and its process for billing for costs incurred. Furthermore, we reviewed publicly available claims reports from BP and GCCF for claims amounts paid. We did not test the claims data or amounts reported by BP or GCCF.

We conducted this performance audit from July 2010 to October 2010 in accordance with generally accepted government auditing standards. Those standards require that we plan and perform the audit to obtain sufficient, appropriate evidence to provide a reasonable basis for our findings and conclusions based on our audit objectives. We believe that the evidence obtained provides a reasonable basis for our findings and conclusions based on our audit objectives.

Liability for Costs Associated with Oil Spills

A complex landscape of laws and regulations governs liability for coastal oil spill costs. The application of different laws and regulations affects which parties ultimately are subject to the various liabilities and financial risks.

In 1990, after the *Exxon Valdez* oil spill in the Prince William Sound in Alaska, Congress enacted the OPA which established a "polluter pays" system that generally places the primary liability for the cost of oil spills on the facility or vessel[44] owner or operator or other "responsible party,"[45] subject to certain limitations on the extent and amount of liability. Under this system, OPA allocates liability to the responsible party for certain spill costs, which include, up to specified limits where applicable, *removal costs* (cleaning up the spill) and *damage claims* (such as restoring the environment and payment of compensation to parties who were economically harmed by the spill) and which are sometimes referred to as "OPA-compensable costs." To pay specified costs above a liability limit, as well as to pay costs when a responsible party does not pay or cannot be identified, OPA authorizes use of the Fund. The Fund is not available for certain types of injuries or damages that may arise in an oil spill incident, such as personal injuries or death or financial losses associated with oil company investments by members of the public. Such damages and injuries may be governed by other federal statutes, common law, or various state laws.

The federal government's potential fiscal exposure increases as responsible parties' capacity to cover their exposures is reduced, whether as a result of extensive oil spill liabilities, poor financial condition of responsible parties, or other factors. OPA and other federal laws[46] set out the federal process for oil spill response management, the liability of different parties in the event of an oil spill incident, and federal funding to cover certain oil spill costs that are not covered by private parties engaged in maritime activities.[47]

The *National Contingency Plan* (NCP), 40 C.F.R. part 300, generally sets out the organizational structure and procedures for preparing for and responding to discharges of oil and releases of hazardous substances, pollutants, and contaminants. This enclosure provides a summary of this framework as applied to oil spills in coastal waters.[48]

Oil Spill Response Management Process

The federal framework includes a coordinated response process involving private parties and all levels of government. In the event of the discharge of oil (or hazardous substances) in coastal waters, any person in charge of a vessel or facility must give notice to the National Response Center,[49] which is administered by the Coast Guard as part of the NCP.[50] Responses to large oil spills are typically a cooperative effort between the public and private sector, and there are numerous players who participate in responding to and paying for oil spills. To manage the response effort, the responsible party, the Coast Guard, EPA, and the pertinent state and local agencies form the unified command , also known as the National Response System, which implements and manages the spill response.[51] Responses can include activities undertaken by federal agencies, states, tribal and local governments, responsible parties (e.g., owners and operators of vessels or facilities) and their guarantors (i.e., insurers), and nongovernmental participants, including qualified individuals[52] and Oil Spill Response Organizations, which usually s erve as contractors to the federal government or resp onsible parties to support response activities, such as skimming or disposing of oil.

The lead federal authority, or Federal On-Scene Coordinator (FOSC), in conducting a spill response is usually the nearest Coast Guard Sector and is headed by the Coast Guard Captain of the Port.[53] Under the NCP, when the Coast Guard designates an oil spill as a spill of national significance, the Coast Guard may appoint a National Incident Commander to provide strategic-level coordination among all public and private international and domestic parties.[54] When notice of an oil spill is received by the Coast Guard, and as soon as the source is identified, NPFC must notify the responsible party or parties by telephone, telefax, or other means of NPFC's designation.[55]

According to NPFC guidance, "the [responsible party] has primary responsibility for response to a spill incident, including setting up the [Incident Command System] and joining with the FOSC and state on-scene coordinator (SOSC) in the [unified command]."[56] However, as reflected in the NCP, NPFC guidance explains that, "even when the responsible party leads a reasonable response effort, the FOSC is always in ultimate command and may decide to direct specific action or, for whatever reason it is deemed necessary, actually take the lead role in the response." Under the Clean Water Act and the Comprehensive Environmental Liability and Compensation Act (CERCLA), and as set out in the NCP, the FOSC may issue administrative orders to protect the public health or welfare to responsible p arties and other private parties directing them to take action to assist in a response.[57]

The FOSC directs response efforts at the scene of an oil spill.[58] Additionally, to obtain assistance from other government agencies, the FOSC issues pollution removal funding authorizations—commitments that the agency will receive reimbursement from the Fund for performing response activities—to obtain services and assistance from other government agencies. Other federal agencies may also be involved. For example, NOAA provides scientific support, monitors and predicts the movement of oil, and conducts environmental analyses of the affected area. The federal, state, and tribal trustees perform one or more natural resource damage assessments, as appropriate. Additionally, regional governmental entities that are affected by the spill—both state and local—as well as tribal government officials or representatives may participate in the unified command and contribute to the response effort, and they may be entitled to reimbursement from the responsible party or the Fund.

Oil Spill Liability and Federal Funding

A complex landscape of laws and regulations governs the liability of various parties for coastal oil spill costs. The applicability of different laws and regulations affects which parties ultimately are subject to the various liabilities and financial risks.

Overview of Liability for Oil Spills

OPA, which among other things amended the CWA, identifies the types of parties liab le for costs and damages it covers, known as "responsible parties." OPA requires that, subject to certain exceptions, all nonfederal claims for OPA-compensable removal costs or damages be submitted first to the responsible party or the responsible party's guarantor.[59] As discussed further below, federal agencies are authorized to use the Fund to cover their removal costs, and the federal government is entitled to reimbursement from responsible parties. Under OPA, responsible parties for vessels, [60] are the owners, operators, or charterers.[61] Responsible parties for offshore facilities are those parties that hold leases or permits for lands on the outer continental shelf.[62] Independent third parties can be liable under OPA if they are the sole cause of the oil spill, and others who insure responsible parties (guarantors) are also liable jointly with the responsible parties (up to the amount of their guarantee).[63]

Table 2. Possible Liability for an Oil Spill in Coastal Waters

Type of harm or penalty	Selected parties who may be entitled to recover	Applicable laws
Removal costs (or cleanup costs)		
• Removal of oil or hazardous substances	Federal government;a states; Indian tribes; and any person taking action consistent with the National Contingency Plan	OPA, CERCLA, and state laws
• Disposal	Federal government; states; Indian tribes; and any person taking action consistent with the National Contingency Plan	OPA, CERCLA, and state laws
• Personnel	Federal government; states	OPA, CERCLA, and state laws
• Prevention of further spills	Federal government; states; Indian tribes; and any person taking action consistent with the National Contingency Plan	OPA, CERCLA, and state laws
Damages		
• Natural resource damages	Federal, state, and foreign governments and Indian tribes	OPA, CERCLA, and state laws
• Real or personal property damage	Federal, state, local and foreign governments, Indian tribes , and private organ izations and individuals	OPA and state laws
• Subsistenc damages e use	Indian tribes individuals , private organizations, individuals	OPA and state laws
• Loss of earning capacity	Companies or individuals with loss of profits or income (One does not have to own the damaged prop erty or resources to submi t a claim under this category.)	OPA and state laws
• Loss of government revenues or profit s	Federal agenc ies; states; local governments	OPA and state laws
• Increased pub lic service costs	States and local governments	OPA and state laws
• Lost investments due to false or misleading statements about oil-related activities leading to share inflation by companies	Shareholders	Federal securities laws
Injuries to the person (including death)		

Table 2. (Continued)

Type of harm or penalty	Selected parties who may be entitled to recover	Applicable laws
• Physical injuries to or death of workers in the production, transportation , or storage of oil	Depending upon applicable laws, the injured seaman, or personal representatives of the deceased in case of death; family members of the deceased; the injured mar itime worker or family members of deceased.	Jones Act, Death on the High Seas Act (DOHSA), Longshore and Habor Workers Compensation Act (LHWCA), Outer Continental Shelf La nds Act (OCSLA).[b]
• Physical harm to or death of bystanders, including m embers of the public	Individuals suffering the injury	Admiralty and maritime tort law; state tort laws
• Emotional distress		
• Pain and suffering		
Criminal an d Civil Fines and Penalties		
• Civil Damages or C Judicial Penalties ivil	Federal and state governments	Clean Water Ac and state laws t (CWA), OCSLA
• Criminal Fin es and Penaltiesc	Federal and state governments	CWA, Endangered Speci (ESA), Marine Mammal Protection Act (MMPA), Migratory Bird Treaty Act (MBTA), OCSLA, Occupational Safety, Health Act (OSH laws es Act A), Refuse Act, and state
• Civil Administrative Fines and Penalties	Federal and state governments	CWA, ESA, MMPA, OPA, OSHA, and state laws

Source: GAO.

[a].At least one U.S. district court has held that the federal government's right to recover cleanup costs is not unlimited under OPA. *United States v. John Paul Jones, Jr.*, 267 F. Supp. 2d 1349, 1363-64 (M.D. Ga. 2003). The court ruled that the federal government may take steps to recover its cleanup costs except to the extent that the responding party establishes that those costs were incurred in an arbitrary and capricious manner. The court explained that federal courts have the authority to review and set aside any government "action" in cleaning up the oil spill under the "arbitrary, capricious, an abuse of discretion, or otherwise not in accordance with law" standard in the Administrative Procedures Act, 5 U.S.C. § 706(2)(A).

[b].Applicable provisions of federal laws include: Jones Act, 46 U.S.C. § 30104; Death on the High Seas Act (DOHSA), 46 U.S.C. App. § 761; Longshore and Harbor Workers Compensation Act (LHWCA), 33 U.S.C. § 901; Outer Continental Shelf Lands Act (OCSLA), 43 U.S.C. §§ 1331-1356a, as amended.

c .Miscellaneous Provisions of Title 18 of the U.S. Code (federal penal code) are often used in the prosecution of federal environmental crimes, such as Aiding and Abetting, 18 U.S.C. § 2; Conspiracy, 18 U.S.C. § 371; False Statements, 18 U.S.C. § 1001; Mail Fraud, 18 U.S.C. § 1341; Wire Fraud, 18 U.S.C. § 1 343; Obstruction of Justice, 18 U.S.C. § 1512; the Criminal Fine Improvements Act, 18 U.S.C. § 3571.

OPA addresses the discharge of oil from a facility or vessel as those terms are defined in the act.[64] Responses and liability for hazardous substances not covered by OPA, which can occur in connection with an oil spill, are addressed under CERCLA.[65] Also, OPA does not apply to oil spills from a public vessel, which OPA defines as "a vessel owne d or bareboat chartered and operated by the United States, or by a State or political subdivision thereof, or by a foreign nation, except when the vessel is engaged in commerce. "[66]

Congress expressly reserved for each state the authority for it to impose additional liability or requirements related to oil spills that occur within the state, including the state's coastal waters.[67] Finally, state oil pollution laws as well as other federal and state laws establish liability for a number of damages, injuries, costs, and fines and penalties not covered by OPA. For example, liability for physical and emotional harms resulting from an oil spill is established under federal maritime and state tort laws depending on various factors, such as where the incident or harm occurs.

Table 2 lists a variety of damages, injuries, cleanup costs, and fines and penalties that can be incurred in relation to an oil spill incident in coastal waters and identifies the parties that generally are entitled to recover financially from responsible parties, assuming that relevant legal criteria have been met. Depending on the circumstances surrounding any particular oil spill incident, a party may incur liability for any combination of these damages, injuries, cleanup costs, and fines and penalties.

OPA requires that, subject to certain exceptions, all claims by individuals, businesses, states, and local and tribal governments for OPA-compensable removal costs or damage s be submitted first to the responsible party or the responsible party's guarantor. Unless NPFC's designation of a responsible party for an oil spill is denied, or if the source of the oil spill is either a public vessel, or unidentifiable, NPFC will instruct the responsible party on the means of advertising the claims procedure to the public and other interested parties.[68] If the responsible party denies a claim or does not settle it within 90 days, a claimant may commence action in court against the responsible party, or present the claim to the NPFC. Again, this process applies only to the costs and damages for which liability arises under OPA.

Under OPA, certain parties must demonstrate financial responsibility before engaging in activities that may result in financial liability under the act. OPA requires the responsible parties for vessels and off-shore facilities to demonstrate the financial resources to respond to an oil spill up to the statutory limit of liability. Under various implementing regulations, the processes and applicable oversight by federal and state agencies for demonstratin g financial responsibility vary according to the different facilities or vessels being used.

For example, before any vessel or mobile offshore drilling unit larger than 300 gross tons operates in United States waters or any vessel that transships or transfers oil in the Exclusive Economic Zone, the responsible parties for oil spills under OPA must first obtain a Certificate of Financial Responsibility (CoFR) from NPFC.[69] To receive a C oFR, the responsible party must demonstrate adequate financial resources to cover the maximum potential liability for an oil spill from a vessel, up to the limits under OPA. Responsible parties may self-insure their financial responsibility requirements or they may purchase insurance from a guarantor, which NPFC must also certify as being contractually obligated to cover the responsible party's maximum potential OPA liability. Failure to comply with the C oFR requirements can result in denial of authority to operate in United States waters.[70]

Users of offshore facilities on outer continental shelf lands have similar requirements to those covering vessels. Under OPA, they must submit evidence of Oil Spill Financial Responsibility (OSFR) for an offshore facility (that generally is capable of discharging more than 1,000 barrels of oil) to the Department of the Interior (Interior) and receive its approval.[71] In implementing these requirements, Interior's Bureau of Ocean Energy Management, Regulation, and Enforcement (BOEMRE) requires potential operators of offshore facilities to obtain approval of their OSFR as a condition of engaging in oil exploration, drilling, or production.[72] Under OPA, the holder of the permit for use of the outer continental shelf lands is a responsible party for liabilities related to oil spills from an offshore facility,[73] but if the operator of the offshore facility is not the permit holder, BOEMRE requires the operator to contractually agree to joint liability. As with vessels, offshore facility operators or permit holders may self-insure their financial responsibility requirements or they may purchase insurance from a guarantor.

These OPA requirements for demonstrating financial responsibility apply only to the statutory maximum amount of potential liability under OPA.[74] It should be noted that OPA places limits on the amount of liability of responsible parties, unless the oil discharge is the result of gross negligence or willful misconduct, or a violation of federal operation, safety, and construction regulations, in which case liability is unlimited.[75] A limitation of liability also is not available where the responsible party fails or refuses to: report the incident as required by law, provide all reasonable cooperation and assistance reque sted by a responsible official regarding removal activities or without sufficient cause, to comply with an order issued under subsection (c) or (e) of 33 U.S.C. § 1321 of 33 U.S.C. § 1 471.

Table 3. Description of Vessels and Offshore Facilities and Current Limits of Liability

Vessels or Facilities	Description	Limit of liability
Oil tanker	An oil tanker is a ship designed to carry oil in large tanks.	Single hull: Vessels greater than 3,000 gross tons: the greater of $3,200 per gross ton or $23,496,000.
		Vessels less than or equal to 3,000 gross tons: the greater of $3,200 per gross ton or $6,408,000.
		Double hull: Vessels greater than 3,000 gross tons: the greater of $2,000 per gross ton or $17,088,000.
Tank barge	A tank barge is a non-self-propelled vessel that carries liquid, solid, or gaseous cargos in bulk in tanks primarily through rivers and inland waterways.	Vessels less than or equal to 3,000 gross tons: the greater of $2,000 per gross on or $4,272,000.
Cargo ship or freighter	A cargo ship or freighter is a vessel That transports non-oil goods and materials.	The greater of $1,000 000 per gross ton or $854,400.
Fishing vessel	A fishing vessel is a ship that is used to catch fish for commercial use.	
Offshore facility	An offshore facility is any facility of any kind located in, on, or under any of the navigable waters of the U.S., and any facility of any kind that is subject to the jurisdiction of the U.S. and is located in, on, or under any other waters, other than a vessel or a public vessel.	All cleanup costs plus $75 million.
Mobile offshore drilling unit (MODU)	A MODU is a vessel (other than a self-elevating lift vessel) capable of use as an offshore facility.	For a discharge on or above the surface of the water, a MODU is first treated as a tank vessel up to the limit of liability for tank vessels. For costs above the vessel liability limit, the MODU is treated as an offshore facility.

Source: GAO.

In general, liability limits under the OPA depend on the kind of vessel or facility from which a spill originates.[76] For an offshore facility like the Macondo Prospect well, statutory liability is limited to all removal costs plus $75 million for damages. For tank vessels, liability limits are based on the vessel's tonnage and hull type, although a statutory division of liability for mobile offshore drilling units results in a range of possible liability limits. For spills on or above the surface of the water, mobile offshore drilling units are treated first as tank vessels up to the limit of liability for tank vessels and then as offshore facilities. The current liability limits under OPA are set out in Table 3.

OPA requires the Coast Guard to adjust the limits of liability not less often than every 3 years to reflect significant increases in the Consumer Price Index, but as GAO recently pointed out, further attention is needed to adjust the liability limits. *See* GAO-10-795T and GAO-07-1085.

Oil Spill Liability Trust Fund

Congress created the Fund[77] to finance the removal costs and damages covered by OPA and the CWA to the extent that responsible parties invoke defenses to or limits on their liability, are unable or unwilling to pay, or cannot be identified.[78] OPA provides that the Fund is available to pay for OPA-compensable removal costs and damages, including federal agencies' costs.[79] Accordingly, OPA requires the President to issue regulations governing how federal officials incur obligations for OPA-compensable costs.[80] In general, the President charged the Coast Guard with administering the Fund, but also authorized federal agencies that incur oil spill response costs and related costs to establish their own policies and procedures for doing so.[81] Further, a state may obligate up to $250,000 from the Fund for the costs of immediate removal of oil or the mitigation or prevention of a discharge, pursuant to a request approved by or a written agreement entered into with the federal government.[82]

Under NPFC policy, the FOSC controls the use of the Fund for removal costs. First, the FOSC coordinates with federal agencies involved in a response to charge certain costs directly against the Fund. Second, the FOSC authorizes federal agencies to incur response costs chargeable to the Fund under a Pollution Removal Funding Authorization (PRFA) or Military Interdepartmental Purchase Request (MIPR). Federal agencies coordinate with NPFC to document these removal costs. NPFC issues guidance to federal agencies on appropriate accounting and related procedures to support these costs.[83]

Under OPA, and pursuant to the Energy Policy Act of 2005 and the Energy Improvement and Extension Act of 2008, the Fund's revenue is generated by a per barrel tax of eight cents on petroleum products received at a United States refinery or imported from other countries. The barrel tax is set to expire in 2017 (after an increase to nine cents per barrel for that year). The fund also receives reimbursements from responsible parties for response costs and damage claims paid by the Fund. For example, as of October 19, 2010, BP had reimbursed the Fund for $518.4 million for *Deepwater Horizon* cleanup costs incurred by Federal agencies. Other funds that flow into the Fund are recoveries of penalties paid pursuant to various statutes, including the Clean Water Act and Deepwater Port Act; and interest earned on U.S. Treasury investments.[84]

OPA authorizes the use of the Fund for, among other specified purposes:[85]

- Removal costs and related monitoring activities that are consistent with the NCP;
- Natural resource damages include the costs to assess the natural resources, and the costs to restore, rehabilitate, replace or acquire equivalent natural resources;
- Payment of claims submitted by private claimants, and states, local, and Indian government claimants; and
- Payment of federal administrative, operational, and personnel costs and other expenses to implement OPA and the Clean Water Act, provided:
- Not more than $25 million shall be available per fiscal year for "the operating expenses incurred by the Coast Guard;"[86] and
- Not more than $27.25 million per year shall be available for oil pollution research.[87]

Expenditures may be made out of the Fund only pursuant to appropriations acts, except for the following:[88]

- Reimbursements for natural resource damages (33 U.S.C. § 2706(f));
- Payment of claims to claimants (33 U.S.C. § 2712(a)(4));
- Expenses related to the Oil Spill Recovery Institute (33 U.S.C. §§ 2731-2736);
- Natural resource damage assessments (33 U.S.C. § 2706);
- Payment of up to $50 million per fiscal year for immediate oil spill removal costs under the Clean Water Act (33 U.S.C. §§ 1321(c), 2752(b)); and

- Emergency advances for any oil spill, up to $100 million, or specifically for the *Deepwater Horizon* incident an unlimited amount (in increments of up to $100 million), subject to overall limitations on the use of the Fund (see below).[89]

The Fund is divided administratively into a Principal Fund and an Emergency Fund. The Emergency Fund accounts for all amounts designated for immediate oil spill removal costs (up to $50 million) and all emergency advances (increments up to $100 million) for an oil spill incident. All other amounts are accounted for in the Principal Fund. See Figure 4 for a depiction of the Fund's structure.

OPA established limits on federal expenditures from the Fund, currently set at the lesser of either:

- The balance of the Principal Fund, including reimbursements from responsible parties and up to $1 billion that may be borrowed from the general fund of the U.S. Treasury pursuant to an appropriation act (less a required $30 million minimum balance in the fund), or
- $1 billion in total expenditures per incident and $500 million for natural resource damage assessments and claims per incident.[90]

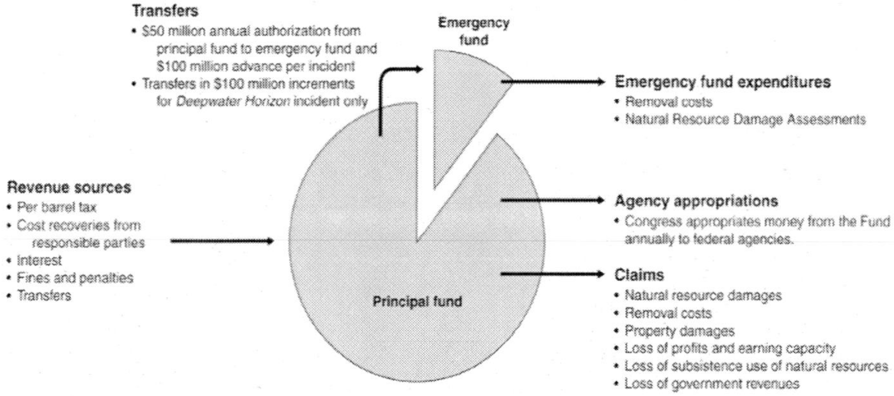

Sources: NPFC and DHS Coast Guard report on implementation of the Oil Polution Act of 1990.

Figure 4. Oil Spill Liability Trust Fund.

Finally, the Fund cannot be used to pay for costs, damages, or injuries that are not covered by OPA and the Clean Water Act.[91] Certain response costs for discharges of hazardous substances that are covered by CERCLA are funded out of the Superfund.[92] Likewise, certain response costs may be funded out of the Disaster Relief Fund if the President makes a disaster declaration. (Such a declaration was not made in response to the *Deepwater Horizon* incident.)

ENCLOSURE IV: DEEPWATER HORIZON OIL SPILL

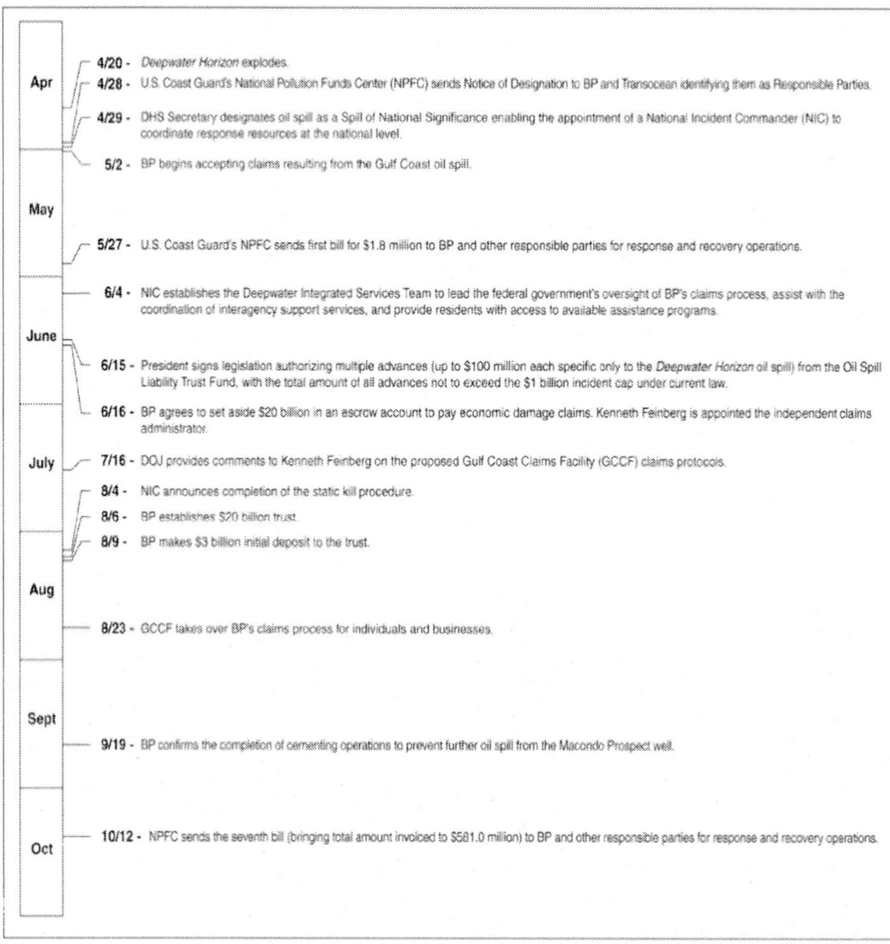

Source: GAO analysis of key events for the Deepwater Horizon Oil Spill.

Figure 5. Timeline of *Deepwater Horizon* Oil Spill Events.

Enclosure V: Comments from the Department of Homeland Security

General Comments

The Coast Guard's authorities and policies in response to the many oil spills that occur in the United States on a day-to-day basis are fundamentally sound and effective. Deepwater Horizon (DWH), the first ever Spill of National Significance (SONS), is highlighting certain authority and policy refinements the National Pollution Fund Center should consider adopting for use in a SONS or other extraordinary events. The Oil Spill Liability Trust Fund is available only to the extent that costs are allowed by the Oil Pollution Act. Not all costs resulting from the Deepwater Horizon oil spill are OPA related costs.

Billing Process

Since 1991, Standard Operating Procedure to fund the interagency cost to an OPA event involves access to the OSLTF using interagency agreements (PRFA and MIPR) that result in obligations. Agency participation culminates in submission of all cost documentation supporting agency expenditures and a request for reimbursement. NPFC reimbursement becomes the actual expenditure and basis of any cost recovery against the RP. NPFC's standard billing process requires completion of response activities and a full reconciliation of all related costs. Due to the unprecedented nature of the Deepwater Horizon response, the NPFC, the federal agencies involved in the DWH response, and BP collectively agreed to a billing procedure in which the NPFC would provide BP a bill consolidating major Federal and state government response obligations on a periodic basis. The consolidated periodic bill requested BP remit 75% of the major Federal and state obligations listed. In return, BP agreed to remit the 75% of the obligated amount immediately to allow the government agencies to continue their part of the response efforts. The utility and need for the billing strategy in use for DWH will be evaluated as part of the lessons learned from this event for use in future catastrophic incidents. The Federal Agencies' completion of response operations necessarily follows in time after the initial obligation of funds, as does the collection and submission of costs to support reimbursement. DHS/CG Chief Financial Officers (CFOs) are working closely with other Federal Agency CFOs to expedite this reimbursement process. We have no reason to believe actual costs will vary substantially from these estimates. The fact that 38.8% of the amounts obligated by Federal Agencies were reimbursed from the OSLTF as 9/30/2010 is in large part due to the fact that the bills accounted for obligations – rather than past expenses. Agencies continue to submit paperwork to support reimbursement on a rolling basis. For all activities that were approved in a MIPR or a PRFA, agencies are reimbursed out of the OSLTF at the time when they submit the paperwork, as would be standard for any reimbursable agency expense. DHS/CG Chief Financial Officers (CFOs) are working closely with other Federal Agency CFOs to help them expedite the reimbursement process.

The Per Incident Cap

In respect to GAO draft conclusions regarding the $1 billion per incident cap on Fund payments, we agree that enacting new legislation concerning the cap will be essential to ensuring established policies and procedures are fully and effectively implemented. In May 2010, the Administration asked Congress to increase the per-incident limit to $1.5 billion. We appreciate GAO's input and look forward to working with Congress to address this important issue.

Gulf Coast Claims Facility

Designation was accepted and BP, one of the responsible parties (RPs), initially established and advertised a claims process. Subsequently, the Administration and BP mutually agreed that Kenneth Feinberg would run an independent claims process. Feinberg established the Gulf Coast Claims Facility, which took over claims operations on August 23, 2010. Claims previously filed with the BP claims process were transitioned to the GCCF for review, evaluation, and determination. Claimants have subsequently filed new claims with the GCCF. If an RP or the GCCF does not satisfy a claimant within 90 days of submission or if they deny the claim, that claimant may elect to pursue the claim in court or to submit the claim to NPFC for consideration of reimbursement from the OSLTF. The NPFC claims process applies the requirements of OPA and the implementing claims regulations in independently arrive at their own determination of compensability of claims. Reimbursement from RPs would be pursued for any claims compensated from the OSLTF. BP also agreed to establish a $20 billion trust to ensure that funds would be available for, among other things, claims designated for payment by the GCCF. The trust is administered by independent trustees.

Integrated Services Team

The Integrated Services Team (IST) was established as a part of the National Incident Command (NIC) on June 4, 2010 to coordinate the efforts of Federal Departments and Agencies on issues related to supportive services to individuals and local governments as well as claims monitoring. The IST had field-based teams stationed in each affected state to coordinate the IST's work in support of the State. Federal Resource Coordinators (FRC's) assigned to each state ensured coordination of information, issue resolution, and service delivery to those impacted by the spill. The IST did not focus on the response and cleanup efforts being directed by the NIC. The IST coordinated interagency and intergovernmental efforts to independently monitor the status of claims submitted to the RPs and, once it was established, the GCCF, and the effectiveness and efficiency of their claims processes. The team took steps to raise awareness of concerns related to payment policy clarity for claimants, process efficiency and effectiveness, improved communication with claimants, data access and reporting, and coordination of federal/state benefits and services to avoid duplicate payments. The IST also hosted conference calls to facilitate dialogue and information sharing between GCCF, BP and federal, state and local government officials. In conjunction with the stand-down of the NIC on September 30, IST capability has been scaled back, with reduced IST capability expected to continue through December 2010 for claims monitoring and coordination. Eventually, responsibility for coordination will be transitioned to other federal agencies under inherent authorities as appropriate. Formal claims monitoring functions will not go to the USCG National Pollution Funds Center.

Notice of Designation

There is confusion about the purpose and impact of a notice of designation. A RP is liable if he owns or operates a facility or vessel from which oil discharges or poses a substantial threat of discharge to waters of the U.S. Whether the RP accepts or denies designation does not affect ultimate liability. The purpose of designation is to get the RP to advertise the claims process as quickly as possible. If the RP denies designation the NPFC will advertise and will continue to pursue any RP for removal costs and damages subject to any OPA defense or limit of liability.

End Notes

[1] BP America Production Company, a subsidiary of BP p.l.c., leased the *Deepwater Horizon* from Transocean Holdings LLC, a subsidiary of Transocean Limited. Transocean Limited is the world's largest offshore drilling contractor comprising numerous subsidiaries and jointly controlled entities and associates. Unless otherwise referring to specific subsidiaries or affiliates, we refer to Transocean Limited and its components separately or jointly as "Transocean." BP p.l.c. is an international oil and gas company comprising numerous subsidiaries and jointly controlled entities and associates. Unless otherwise referring to specific subsidiaries or affiliates, we refer to BP p.l.c. and its components separately or jointly as "BP." BP was originally incorporated in 1909 in England and Wales as "British Petroleum" and changed its name in 2001.

[2] BP established the trust under Delaware law, which generally provides that the principal of the trust can be used only for the purposes stated in the trust agreement and that the terms of the trust agreement cannot be modified and are legally enforceable by the trustees.

[3] BP, *BP p.l.c. Group results, second quarter and half year 2010*, London, July 27, 2010 and Oxford Economics, *Potential Impact of the Gulf Oil Spill on Tourism*, July 22, 2010.

[4] Pub. L. No. 101-380, 104 Stat. 489 (1990).

[5] The Fund also pays for the costs of certain federal agency operations.

[6] This work focused on the *Deepwater Horizon* oil spill. Our past work identified a number of other risks and vulnerabilities to the fund. For example, the Fund is at risk from claims resulting from spills that significantly exceed responsible parties' liability limits. See GAO, *Maritime Transportation: Major Oil Spills Occur Infrequently, but Risks to the Federal Oil Spill Fund Remain*, GAO-07-1085 (Washington, D.C.: Sept. 7, 2007).

[7] The $1 billion per incident cap applies concurrently with a $500 million per incident cap for natural resource damages and related assessment costs. The administration submitted a legislative proposal to Congress in May 2010 to increase the caps to $1.5 billion and $750 million per incident, respectively, to address the effects of inflation since the caps were enacted in 1990.

[8] If, as for the *Deepwater Horizon* oil spill, the Fund is being fully reimbursed and net expenditures are zero, the cap still applies.

[9] An obligation is a commitment, such as a contract, that creates a legal liability for the payment of goods and services ordered or received.

[10] The NCP is a national plan maintained by the Environmental Protection Agency (EPA) that provides the organizational structure and procedures for preparing for and responding to discharges of oil and releases of hazardous substances, pollutants, and contaminants. Appendix E to the NCP provides the organizational structure and procedures to prepare for and respond to oil spills.

[11] An oil spill is deemed to be of national significance if "due to its severity, size, location, actual or potential impact on the public health and welfare or the environment, or the necessary response effort, is so complex that it requires extraordinary coordination of federal, state, local, and responsible party resources to contain and clean up the discharge."

[12] Pub. L. No. 101-380, 104 Stat. 489 (1990).

[13] Enclosure III presents the liability limits.

[14] In 2017, the per barrel tax increases to 9 cents; however, the tax is scheduled to expire at the end of 2017.

[15] The Coast Guard has responsibility for removal actions in the coastal zone, while EPA has responsibility in the inland zone. Because the *Deepwater Horizon* oil spill occurred in the Gulf of Mexico, the Coast Guard is the Federal On-Scene Coordinator for the incident.

[16] Oil Spill Response Organizations (OSROs) are companies that specialize in cleaning up oil spills. They often serve as contractors or subcontractors for spill response efforts.

[17] Federal authorizations authorize reimbursement of federal and nonfederal government agencies from the Fund for oil response and removal activities. NPFC uses MIPRs rather than federal authorizations for the Department of Defense, to authorize reimbursement.

[18] On May 11, 2010, NPFC notified BP and Transocean Holdings Incorporated that BP's advertising and claims processing were sufficient, and Transocean should not advertise and should coordinate claims processing with BP. According to NPFC officials, they wanted to avoid public confusion and have only one Responsible Party advertise for claims.

[19] The three trustees are Citigroup Trust-Delaware, N.A., which serves as the corporate trustee, and John S. Martin, Jr. and Kent D. Syverud, who serve as the individual trustees.

[20] The funding schedule for the escrow account agreed to by the administration and BP was for contributions by BP of $5 billon a year for 4 years. BP later confirmed that the funding schedule would include an initial deposit of $3 billion, which was made on August 9, 2010, with an additional deposit of $2 billion in the fourth quarter of 2010 and $1.25 billion a quarter until the entire $20 billion has been deposited.

[21] The Jones Act 46 U.S.C. § 30104, establishes liability for injury or death of seamen incurred in the course of their employment.

[22] See http://www.gulfcoastclaimsfacility.com/ (as of October 4, 2010, GCCF had not yet issued protocols for final claim payments).

[23] GAO, *Maritime Transportation: Major Oil Spills Occur Infrequently, but Risks to the Federal Oil Spill Fund Remain*, GAO-07-1085 (Washington, D.C.: Sept. 7, 2007).

[24] BP, *BP p.l.c. Group results, second quarter and half year 2010*, London, July 27, 2010.

[25] Whether the *Deepwater Horizon* oil spill arose as the result of gross negligence, willful misconduct, or a violation of federal operation, safety, or construction regulations remains subject to investigation and potential litigation.

[26] The cap is for total expenditures. If, as for the *Deepwater Horizon* oil spill, the Fund is being fully reimbursed and net expenditures are zero, the cap still applies. [27]The $1 billion cap is concurrent with a $500 million cap on expenditures for natural resource damages and related assessments. It is not just the federal government that is at risk as a result of the cap. In addition, no claims may be paid for private parties' or states' damages and removal costs.

[28] According to the agency, the Coast Guard has historically viewed its OPA recoverable costs as activities normally funded through the agency's operating expense appropriation, and thus it has not sought reimbursement for these costs from the Fund.

[29] Amounts recovered from the Responsible Parties for Coast Guard recoverable costs are deposited into the Fund.

[30] According to NPFC officials, any amounts paid to the Natural Resource Trustee will be billed to the Responsible Parties.

[31] See http://www.bp.com.

[32] According to GCCF's Web site, emergency payments are payments available to individuals and businesses that are experiencing financial hardship resulting from damages incurred by the *Deepwater Horizon* oil spill. Final payments will be issued for claims that individuals and business have filed for income, damages, and other losses associated with the *Deepwater Horizon* oil spill.

[33] An obligation is a commitment, such as a contract, that creates a legal liability for the payment of goods and services ordered or received.

[34] The process, prior to the *Deepwater Horizon* oil spill, was to bill the responsible party based only on actual costs.

[35]NPFC uses Department of Interior's Bureau of Ocean Energy Management, Regulation and Enforcement's leases to identify some Responsible Parties.

[36] NPFC sent the notice to Transocean Holdings Incorporated, but Transocean replied that the correct entity is Transocean Holdings LLC. Further, Transocean asserted to NPFC that it was a responsible party solely for purposes of any oil that was discharged from the mobile offshore drilling unit *Deepwater Horizon* and not from the Macondo Prospect well.

[37] 33 U.S.C. § 2714(a). Coast Guard regulations authorize NPFC to notify Responsible Parties of their designation through any means, but all designations will be confirmed via written Notice of Designation. 33 C.F.R. § 136.305(a).

[38] GAO, *Standards for Internal Control in the Federal Government*, GAO/AIMD-00-21.3.1 (Washington, D.C.: November 1999).

[39] According to White House and BP press releases, the President of the United States and BP "mutually agreed that Mr. Kenneth Feinberg will run the GCCF." According to BP, Mr. Feinberg was jointly appointed by the President and BP to serve as an independent contractor to BP.

[40] Our future work will look at the stand-down and transition to workgroup and agencies in order to understand the longer-term activities.

[41] GAO, Maritime Transportation: Major Oil Spills Occur Infrequently, but Risks to the Federal Oil Spill Fund Remain, GAO-07-1085 (Washington, D.C.: Sept. 7, 2007).

[42] Department of Homeland Security, *Fiscal Year 2009 Annual Financial Report* (Washington, D.C., Nov. 16, 2009).

[43] Department of Homeland Security Office of Inspector General, *Information Technology Management Letter for the United States Coast Guard Component of the FY 2009 DHS Integrated Audit*, OIG-10-77 (Washington, D.C., Apr. 9, 2010).

[44] OPA contains provisions dealing with facilities, including offshore facilities, onshore facilities, pipelines, and deepwater ports, as well as vessels, including mobile offshore drilling units, tank vessels, barges, and cargo, fishing, and other ships. For definitions of these terms, see 33 U.S.C. § 2701.

[45] For example, for an offshore facility, the responsible party is the permittee or lessee of the area in which the facility is located.

[46] For example, a parallel and similar framework exists for costs associated with the release of hazardous substances, which can occur in connection with an oil spill. The Comprehensive Environmental Response and Compensation Act of 1980, as amended (CERCLA), 42 U.S.C. ch. 103, contains liability, discharge response management, and funding provisions similar to and generally integrated with those described in this enclosure for oil spills.

[47] Federal laws and regulations also provide for advance oil spill prevention and response planning. This is generally reflected in the *National Oil and Hazardous Substances Pollution Contingency Plan* (known as the *National Contingency Plan*), 40 C.F.R. pt. 300, the *National Response Framework*, available at http://www.fema.gov/emergency and the *Integrated Contingency Plan*, 61 Fed. Reg. 28642 (June 5, 1996).

[48] Other laws may also apply and different risks may also arise in oil spills that occur in inland waters, including state and private water, and on land. Although the areas of federal and state government jurisdiction over inland and ocean waters varies by federal and state law, in this enclosure, in general, federal waters include all coastal waters extending seaward 200 nautical miles and all inland navigable waters (*i.e.,* waters over which the federal government has jurisdiction to regulate), and state waters include all waters within or adjoining a state and up to 3 nautical miles from the coastline, although certain states and territories own and can regulate waters up to 9 nautical miles from the coastline. For a description of the nature of federal and state jurisdiction over ocean waters and ownership of the related submerged lands, see *An Ocean Blueprint for the 21st Century*, U.S. Commission on Ocean Policy (Wash., D.C.: Sep. 20, 2004), available at http://oceancommission.gov/ (last visited Nov. 2, 2010).

[49] 33 C.F.R. § 153.203; 40 C.F.R. §§ 110.6, 300.125, and 302.6. The primary function of the National Response Center is to serve as the sole national point of contact for reporting all oil, chemical, radiological, biological, and other discharges into the environment anywhere in the United States and its territories.

[50] The NCP establishes the top priority of an oil spill response as saving human life and the next priority as stabilizing the situation to preclude it from worsening, including the prevention of further spilling that would require additional removal actions and to minimize adverse impact to the environment. 40 C.F.R. § 300.317.

[51] The Incident Command System (ICS) is a standardized response management system that is part of the National Interagency Incident Management System. The ICS is organizationally flexible so that it can expand and contract to accommodate spill responses of various sizes. The ICS typically consists of four sections: operations, planning, logistics, and finance/administration.

[52] As part of vessel oil spill response plans, vessels must designate a "qualified individual" who acts with full authority to obligate private funds required to carry out response activities. The qualified individual acts as a liaison with the lead federal authority and is responsible for activating the incident response plan.

[53] For coastal oil spills, the Coast Guard generally serves as the FOSC and for inland oil spills the Environmental Protection Agency generally serves as the FOSC.

[54] 40 C.F.R. § 300.323(c). Coast Guard Admiral Thad Allen was designated as the National Incident Commander for the *Deepwater Horizon* incident.

[55] 33 U.S.C. § 2714(a); 33 C.F.R. § 136.305(a). A written notice of designation confirms any designation.

[56] NPFC, *NPFC User Reference Guide (eURG)*, Appendix B, *FOSC Funding Information for Oil Spills and Hazardous Materials Releases* (Wash., D.C.: April 2003), *available at* http://uscg.mil/npfc/URG/default.asp (accessed Nov. 4, 2010).

[57] See 40 C.F.R. § 300.130 for applicable procedures and requirements.

[58] On May 22, 2010, the President established a National Commission on the BP Deepwater Horizon Oil Spill and Offshore Drilling to examine the root causes of the *Deepwater Horizon* incident and options for guarding against, and mitigating the impact of, oil spills

associated with offshore drilling. The Commission is evaluating the effectiveness of the NCP response in the *Deepwater Horizon* incident. *See* http://www.oilspillcommission.gov.

[59] 33 U.S.C. § 2713(a).

[60] A mobile offshore drilling unit such as *Deepwater Horizon* is a vessel capable of use as an offshore facility.

[61] 33 U.S.C. § 2701(32)(A).

[62] 33 U.S.C. § 2701(32)(C).

[63] 33 U.S.C. §§ 2702(d), 2716.

[64] Oil spills governed by OPA are discharges of oil into coastal federal and state waters (in addition to inland navigable waters, which are not addressed in this enclosure). A "discharge" of oil covered by OPA includes "any emission (other than natural seepage), intentional or unintentional, and includes, but is not limited to, spilling, leaking, pumping, pouring, emitting, emptying, or dumping." 33 U.S.C. § 2701(7).

[65] *See* section 101(14) of CERCLA, 42 U.S.C. § 9601.

[66] 33 U.S.C. § 2701(29).

[67] 33 U.S.C. § 2718(a) and (b).

[68] 33 U.S.C. § 2714; 33 C.F.R. §§ 136.309, 136.311, and 136.313.

[69] The Coast Guard regulations for CoFR process are set out at 33 C.F.R. part 138.

[70] 33 U.S.C. § 2716(b). A state may enforce these financial responsibility requirements. 33 U.S.C. § 2719.

[71] Interior regulations for the OSFR process are set out in 30 C.F.R. part 253. Interior's Bureau of Ocean Energy Management, Regulation, and Enforcement (BOEMRE), formerly the Minerals Management Service (MMS), administers this process in conjunction with its authority to lease outercontinental shelf lands and oversee oil exploration and production. *See, generally,* 43 U.S.C. §§ 1331-1356a and 30 C.F.R. parts 250 and 260. GAO and the Department of the Interior Office of Inspector General have issued several reports identifying problems in MMS's oil and gas program leasing and revenue collection programs. See GAO-10-888T, *Oil and Gas Management: Past Work Offers Insights to Consider in Restructuring Interior's Oversight* (Wash., D.C.: July 22, 2010) for a discussion of this work.

[72] 30 C.F.R. § 253.11. The procedures for obtaining permits to operate a deepwater port, a specialized facility under OPA, are established by the Department of Transportation under the Deepwater Port Act of 1974, as amended, 33 U.S.C. § 1501, *et seq.*, in 33 C.F.R. part 148. Guidance on this process, including demonstrations of financial responsibility under OPA, is available on the websites of the Maritime Administration and the United States Coast Guard, http://www.marad.dot.gov/portslandingpage/deepwaterportlicensing/deepwaterportlicensing.htm and http://www.uscg.mil/hq/cg5/cg522/cg5225, respectively. The President delegated the authority to issue regulations governing the demonstration of financial responsibility to Interior for offshore facilities and Transportation for deepwater ports in Executive Order No. 12777, Oct. 22, 1991, as amended by Executive Order No. 13286, Feb. 28, 2003.

[73] 33 U.S.C. § 2701(32)(C).

[74] As noted above, states may impose additional liability and requirements related to oil spills in state waters.

[75] Under OPA, a responsible party can also assert a defense to liability if the oil spill was caused solely by, among other things, an "act of God," an "act of war," the acts or omissions of an independent third party (provided certain conditions are satisfied), or any combination of these. 33 U.S.C. § 2703. NPFC guidance acknowledges that terrorism or other criminal acts

may present a defense to liability under OPA. NPFC, *NPFC User Reference Guide (eURG)*, Appendix B, *FOSC Funding Information for Oil Spills and Hazardous Materials Releases* (Wash., D.C.: April 2003), *available at* http://uscg.mil/npfc/URG/default.asp.

[76] 33 U.S.C. § 2704.

[77] Congress created the Fund in 1986, but not until the *Exxon Valdez* grounding and the passage of OPA in 1990, did it authorize use of the Fund.

[78] If the responsible party denies the NPFC designation or cannot pay or if NPFC cannot identify the source or if the source is a public vessel, OPA requires NPFC to advertise the claims process. 33 U.S.C. § 2714(c).

[79] 33 U.S.C. § 2712 and 26 U.S.C. § 9509(e)(1). OPA consolidated the liability and compensation provisions of four prior federal oil pollution initiatives and their respective trust funds into the Oil Spill Liability Trust Fund and authorized the collection of revenue and the use of the money, with certain limitations, with regard to expenditures. The prior federal laws regarding oil pollution were the Federal Water Pollution Control Act, the Deepwater Port Act, the Trans-Alaska Pipeline System Authorization Act, and the Outer Continental Shelf Lands Act Amendments of 1978.

[80] 33 U.S.C. §§ 2712(c), (e).

[81] Executive Order No. 12777, as amended by Executive Order 13286.

[82] 33 U.S.C. § 2712(d).

[83] On July 1, 2010, the Office of Management and Budget issued guidance to federal agencies outlining "sound practices" for tracking all costs incurred in responding to the *Deepwater Horizon* incident, including their direct costs and indirect costs, but specifically stated that it did not supersede NPFC's guidance on identifying and documenting reimbursable removal costs. OMB Memorandum No. M-10-29, *Identifying and Documenting Costs of Government Activities Related to the BP Deepwater Horizon Oil Spill*, July 1, 2010.

[84] 26 U.S.C. § 9509(b). The fund may also borrow up to $1 billion from the general fund of the U.S. Treasury, as may be provided in appropriations acts. 26 U.S.C. § 9509(d).

[85] 26 U.S.C. § 9509(c); 33 U.S.C. § 2712(a).

[86] 33 U.S.C. § 2712(a)(5)(A).

[87] 33 U.S.C. § 2712(a)(5)(C).

[88] 26 U.S.C. § 9509(c); 33 U.S.C. § 2752(a). *See* B-255979, Oct. 30, 1995.

[89] Congress provided these authorities in the Maritime Transportation Security Act of 2002, Pub. L. No. 107- 295, § 323, 116 Stat. 2064, 2104 (2002) and in June and July 2010, in Public Laws 111-191, (Pub. L. No. 111-191, § 1, 124 Stat. 1278 (2010) and 111-212, (Pub. L. No. 111-212, § 2001, 124 Stat. 2302, 2337 (2010), respectively. The Coast Guard must notify Congress of each emergency advance.

[90] 26 U.S.C. § 9509(c)(2). The Department of Homeland Security General Counsel told us that individual federal agencies may receive and use separate appropriations for their response costs and that such costs count toward the Fund's statutory expenditure limits only if and when the Fund reimburses an agency's appropriation. The NCP provides that agencies incurring costs other than oil spill response costs may use their appropriations to the extent that they are available for the specific agency action, such as scientific investigations related to an oil spill. 40 C.F.R. § 300.335(d). Further, the NCP states that the Department of Defense uses certain of its appropriations for specific response activities, such as removal of sunken vessels or other obstructions of navigation. 40 C.F.R. § 300.335(f)(1)(i).

[91] *See* NPFC, *NPFC User Reference Guide (eURG)*, Appendix B, *FOSC Funding Information for Oil Spills and Hazardous Materials Releases* (Wash., D.C.: April 2003), *available at* http://uscg.mil/npfc/URG/default.asp.

[92] *See* 26 U.S.C. § 9507; 42 U.S.C. § 9611.

INDEX

A

B

C

D

E

F

G

H

I

R

S

T

V

W